INTELLIGENT SYSTEMS FOR VIDEO ANALYSIS AND ACCESS OVER THE INTERNET

ISBN 0-13-047117-8

IMSC Press Multimedia Series

ANDREW TESCHER, Series Editor, *Compression Science Corporation*

Advisory Editors
LEONARDO CHIARIGLIONE, *CSELT*
TARIQ S. DURRANI, *University of Strathclyde*
JEFF GRALNICK, *E-splosion Consulting, LLC*
CHRYSOSTOMOS L. "MAX" NIKIAS, *University of Southern California*
ADAM C. POWELL III, *The Freedom Forum*

▶ Desktop Digital Video Production
  Frederic Jones

▶ Touch in Virtual Environments:
  Haptics and the Design of Interactive Systems
  Edited by Margaret L. McLaughlin,
  João P. Hespanha, and Gaurav S. Sukhatme

▶ The MPEG-4 Book
  Edited by Fernando M. B. Pereira and Touradj Ebrahimi

▶ Multimedia Fundamentals, Volume 1:
  Media Coding and Content Processing
  Ralf Steinmetz and Klara Nahrstedt

▶ Intelligent Systems for Video Analysis and Access Over the Internet
  Wensheng Zhou and C.-C. Jay Kuo

The Integrated Media Systems Center (IMSC), a National Science Foundation Engineering Research Center in the University of Southern California's School of Engineering, is a preeminent multimedia and Internet research center. IMSC seeks to develop integrated media systems that dramatically transform the way we work, communicate, learn, teach, and entertain. In an integrated media system, advanced media technologies combine, deliver, and transform information in the form of images, video, audio, animation, graphics, text, and haptics (touch-related technologies). IMSC Press, in partnership with Prentice Hall, publishes cutting-edge research on multimedia and Internet topics. IMSC Press is part of IMSC's educational outreach program.

INTELLIGENT SYSTEMS FOR VIDEO ANALYSIS AND ACCESS OVER THE INTERNET

Wensheng Zhou

C.-C. Jay Kuo

PRENTICE HALL PTR
UPPER SADDLE RIVER, NJ 07458
WWW.PHPTR.COM

Library of Congress Cataloging-in-Publication Data

A catalog record for this book can be obtained from the Library of Congress

Editorial/Production Supervision: *Nick Radhuber*
Acquisitions Editor: *Bernard Goodwin*
Editorial Assistant: *Michelle Vincente*
Marketing Manager: *Dan DePasquale*
Manufacturing Buyer: *Alexis Heydt-Long*
Cover Design: *Talar Boorujy*
Cover Design Director: *Jerry Votta*

 © 2003 by Pearson Education, Inc.
Publishing as Prentice Hall PTR
Upper Saddle River, NJ 07458

Prentice Hall books are widely used by corporations and government agencies for training, marketing, and resale.

The publisher offers discounts on this book when ordered in bulk quantities. For more information, contact Corporate Sales Department, phone: 800-382-3419; fax: 201-236-7141; email: corpsales@prenhall.com Or write: Corporate Sales Department, Prentice Hall PTR, One Lake Street, Upper Saddle River, NJ 07458.

Product and company names mentioned herein are the trademarks or registered trademarks of their respective owners. Figures appearing in Chapters 5 and 6 are reprinted with the permission of CNN, NBC, and the IOC.

All rights reserved. No part of this book may be reproduced, in any form or by any means, without permission in writing from the publisher.

Printed in the United States of America

10 9 8 7 6 5 4 3 2 1

ISBN 0-13-047117-8

Pearson Education LTD.
Pearson Education Australia PTY, Limited
Pearson Education Singapore, Pte. Ltd
Pearson Education North Asia Ltd
Pearson Education Canada, Ltd.
Pearson Educación de Mexico, S.A. de C.V.
Pearson Education—Japan
Pearson Education Malaysia, Pte. Ltd

CONTENTS

PREFACE		xi
1	**INTRODUCTION**	**1**
1.1	Challenges of Multimedia Information Access	2
	1.1.1 Video Database Indexing and Retrieval	3
	1.1.2 Video Summarization and Browsing	3
	1.1.3 Video Streaming and Filtering	4
1.2	Review of Previous Work	5
1.3	Information Dissemination by Multicast	9
	1.3.1 Collaborative Semantic Multicast	9
	1.3.2 On-Line Multimedia Content Analysis and Annotations	12
1.4	System Overview	15
	1.4.1 Layered Video Analysis Model	15
	1.4.2 On-Line Video Content Analysis Infrastructure	18
1.5	Roadmap of This Book	20
1.6	Outline of This Book	24
2	**ON-LINE SCENE CHANGE DETECTION OF (MBONE) MULTICAST VIDEO**	**25**
2.1	Introduction	25
	2.1.1 RTP Protocol	27
	2.1.2 MBone Tools	29
	2.1.3 Vic and Intra-H.261 Algorithm	30
2.2	Related Work	33
2.3	Scene Change Detection Algorithms and Implementation	35
2.4	Experimental Results and Discussion	38
	2.4.1 Results for Algorithms	38
	2.4.2 Discussion	45
2.5	Summary	48

3 VIDEO/AUDIO/TEXT FEATURE REPRESENTATION AND ANALYSIS — 49
- 3.1 Introduction — 49
  - 3.1.1 Clustering in Feature Space — 50
- 3.2 Overall Software Structure — 51
- 3.3 Visual Features Representation and Analysis — 53
  - 3.3.1 Color Descriptors — 53
  - 3.3.2 Edge Descriptors — 60
  - 3.3.3 Shape Descriptors — 61
- 3.4 Motional Features Representation and Analysis — 62
  - 3.4.1 Camera Motion/Motion Flow Clustering Descriptor — 63
  - 3.4.2 Object Motion Descriptor — 64
  - 3.4.3 Motion Activities Descriptor — 65
- 3.5 Audio Features Representation and Analysis — 70
  - 3.5.1 Time-Domain Features — 70
  - 3.5.2 Frequency-Domain Features — 71
- 3.6 Text Feature Representation and Analysis — 73
  - 3.6.1 Frequency-Based Vectors — 74
  - 3.6.2 Boolean Vectors — 75
  - 3.6.3 Dimensionality Reduction Techniques — 75

4 KNOWLEDGE-BASED VIDEO HIERARCHICAL CLASSIFICATION — 78
- 4.1 Introduction — 78
- 4.2 Previous Work — 79
- 4.3 Video Semantic Concept Tree — 82
  - 4.3.1 Characteristics of Video Semantic Content — 82
  - 4.3.2 Creating a Video Semantic Concept Tree — 83
- 4.4 Decision Tree Learning Algorithm — 86
- 4.5 Rule-Based Knowledge Base Construction — 89
  - 4.5.1 Video Classification Rules in the Knowledge Base — 89
  - 4.5.2 Video Classifier with Video/Audio Cues — 92
  - 4.5.3 Video Classifier with Text Cues — 93
- 4.6 On-Line Knowledge-Based Video Classification System — 97
  - 4.6.1 Media Planner (MP) — 97
  - 4.6.2 Media Matcher (MM) — 101
- 4.7 Summary — 103

5 EXPERIMENTAL RESULTS — 104

	5.1	CNN News Video Segmentation and Indexing	104
		5.1.1 TV News Video Production Rules	104
		5.1.2 On-Line CNN News Segmentation Using Scene Classifications	106
		5.1.3 Table of Contents Generation for CNN News Indexing	107
	5.2	Hierarchical Video Classification on NBC 2000 Olympic Video	108
		5.2.1 Program Segmentation of NBC Sports Video	109
		5.2.2 Sports Event Units Segmentation	111
		5.2.3 Low-Level Feature Extraction	111
		5.2.4 Rule-Based Video Event Classification for a Basketball Video	122
		5.2.5 Hierarchical NBC Video Classification	126
	5.3	Comparison with Previous Work	129
	5.4	Summary	131
6	**SYSTEM INTEGRATION**	**132**
	6.1	Introduction	132
	6.2	Semantic Multicast System Architecture	133
	6.3	On-Line Video Content Analysis Implementation	136
		6.3.1 On-Line Scene Change Detection and Key Frame Extraction	137
		6.3.2 Low-Level Feature Analysis and Extraction	139
		6.3.3 Knowledge-Based Video Classification	141
	6.4	Content Agent Coordination	142
		6.4.1 Communication Between the Proxy and the Service Assigner	143
		6.4.2 Concept of the Metadata Channel	144
	6.5	Examples of System Usage	151
		6.5.1 On-Line CNN News Filtering	151
		6.5.2 Off-Line Video Database Query	153
	6.6	Summary	154
7	**CONCLUSION AND FUTURE WORK**	**157**
	7.1	Summary of This Research	157
	7.2	Contributions of This Research	158
	7.3	Future Work	160
REFERENCES			**163**

INDEX 173

PREFACE

The explosion of on-line web information has given rise to many query-based text search engines (such as Alta Vista) and manually constructed topic hierarchies (such as Yahoo!). With the current growth rate of web information, especially broadband multimedia data, query data are growing incomprehensibly large and manual classification in topic hierarchies is creating a major bottleneck. Consequently, the huge amount of multimedia data is imposing on people a heavy burden of manipulating, searching, interpreting, skimming, and integrating information. Thus, efficient multimedia content analysis tools are needed to address these user's needs.

This book presents a solution to problems arising from the demand for fast information access and for sharing in real-time multimedia transmission over the Internet. We present in this book a solution which exploits software agents that are placed throughout the network environment. These hierarchical video analysis agents process multimedia streams in real time, and automatically decompose and understand the multimedia content so as to facilitate information access and sharing.

Multimedia content contains both the perceptual content such as color, motion, or acoustic features and the conceptual content, which is specified based on concepts or semantics that can be expressed by text descriptions. Both types of contents are embedded simultaneously in multimedia streams, and usually are complementary to each other. This book adaptively analyzes both kinds of video contents by combining mixed media cues from audio, video, and text.

First, a high-performance module for on-line video segmentation based on scene-change detection is described. The module serves as the first step of any video stream construction and analysis. To meet the high computational demand, our proposed video scene change detection algorithms are very efficient while maintaining high accuracy and recall rates for fast on-line video analysis.

Second, the perceptual features of audio and video data are analyzed in a bottom-up manner and integrated so as to discriminate among the different events in any video stream effectively. An efficient decision-tree learning algorithm is used to induce a set of if-then rules which link perceptual features with the video conceptual semantic contents. These rules not only serve as a video classifier, but also guide on-line real-time video/audio feature extraction and data redistribution. A novel knowledge-based system, where knowledge is stored as learned rules, is proposed and descibed in this book to serve as a video semantic inference/classification engine.

Third, we present our proposed hierarchical video categorization scheme based on machine learning of the text information contained in a video – a scheme which provides a good complement to the video/audio classification subsystem. The learned text features for each video category are also stored in the knowledge base. To fuse the text classifier and the audio/video classifier, a media cue optimizer that is trained by using the cue probability distribution based on the concept hierarchy is adopted to guide real-time media query and analysis.

The integration of hierarchical video analysis, clustering, and classification allows a large amount of multimedia data to be organized and presented to users in an individualized and comprehensible way. A general hierarchical concept tree scheme is used to organize a video into a table of contents for video applications and enables a comprehensive agent-based solution to real-time multimedia distribution and sharing over the Internet.

ACKNOWLEDGMENTS

Special thanks are due to HRL Laboratories, LLC, and to the program manager, Mr. Son Dao, for his encouragement, understanding, and support on the exciting and challenging research projects. We also want to acknowledge Dr. Jihoon Yang, Dr. Asha Vellaikal, and Dr. Eddie Shek from HRL Laboratories, LLC, for their insightful discussions with us on related research issues, their valuable comments on our research work and, particularly, in the case of Drs. Vellaikal and Shek, for their close work with us on the Semantic Multicast Project, including their input on the initial architecture and problem design, upon which our own video content analysis module was based.

We are grateful to DARPA for providing funding to the Semantic Multicast project under Grant No. N66001-97-2-8901. Some research results presented in this dissertation have been filed as patents under the proprietorship of HRL Labs, LLC.

We would like to thank Mr. Bernard M. Goodwin, the publisher from Prentice Hall PTR, for his incredibly patient help to make this book come true. We would also like to thank Mr. Nicholas Radhuber for assistance with proof corrections, Ms. Lori Hughes for help in solving editing problems and providing editing packages.

Finally we would each like to thank our families for providing excellent encouragement and support throughout the long and seemingly never-ending saga of "the work of the book"!

Chapter 1

INTRODUCTION

New integrated services are emerging from the rapid technological advances in networking, multiagents, media, and broadcasting technologies. This advancement allows for large amounts of multimedia information to be distributed and shared on the Internet. Current information and communication technologies provide the infrastructure to send bits anywhere, but do not presume to handle information at the semantic level due to insufficient indexing mechanisms and the lack of good automated semantic interpretation mechanisms. Consequently, the huge amount of multimedia data imposes on people a heavy burden of data manipulation, including searching/choosing, interpreting, skimming, and integrating information.

As storage and bandwidth capabilities increase, semi-automated or automated understanding of multimedia data becomes increasingly important in various applications, especially when such data has become highly prevalent. Typical applications are multimedia database retrieval, information filtering, content-based navigation, interactive media services, and so on. Traditional video understanding problems, especially as applied to content-based data retrieval, focus on the use of low-level features such as color, motion, and texture for video indexing. While a direct application of generic similarity metrics to low-level features can give good results to those having similar visual features, their applicability to discerning similar semantic classes is highly suspect. This is partly because it is a very difficult task to combine multiple low-level features effectively, and generic models for multiple types of video applications are not feasible.

The need for query with semantic key-words/key-phrases/key-concepts has motivated recent research in semantic video indexing. This is, however, severely hampered by the gap existing between low-level media features and semantics. Besides, video/image semantic contents may not be easily defined since different people often look at the same visual data from different perspectives. For example, the general audience which watches a basketball game may care more about events such as scores, dunks, etc., while a coach may pay more attention to a team's defense/offense strategies. Moreover, semantic information has different levels of detail and granularity. Therefore, there is a high demand to develop efficient mechanisms which can deal with these diverse needs and features.

At the same time, as the amount of on-line multimedia data increases astronomically, the design of an efficient algorithm or an approach to access the data (e.g., classification, clustering, filtering, etc.) has become of great interest too. There are two important aspects to consider for such a design. First, the data need to be arranged efficiently. For example, instead of placing all data in a flat directory, we can arrange them hierarchically based on a concept hierarchy. Examples can be seen in text-searching engines, like Yahoo!, CNN, and other major Internet news directories. Querying with respect to a concept hierarchy is a lot more efficient and reliable than searching for specific keywords since the views of the data collected are refined as we go down the hierarchy. Second, an efficient algorithm for video classification should be used. As videos present very bulky visual data, which are not like text information, it is normally very tedious to index and retrieve these data based on their semantic meanings. Thus automatic video semantic content analysis is very important.

Videos are a multistream media. They contain not only the moving picture signals, but also audio, speech, and text signals. All these media display different characteristics and express information at different levels and with different kinds of details. How to get the most useful semantic contents from mixed media cues presents an interesting and practical problem. In a typical collaboration scenario, there is a presence of multiple types of media such as video, audio, whiteboard data, and so on. While trying to extract semantics from such data, it is necessary to utilize information available from all sources so as to improve the quality of the video content analysis and the created metadata for on-line applications. Semantic content extraction from different media should automatically adapt to sources where there is a meaningful activity.

1.1 Challenges of Multimedia Information Access

As noted previously, users simply cannot manually look through the volumes of multimedia information that are available. Hence, it is becoming clear that computational methods are needed now that help store, query, and filter multimedia information to allow users to readily access documents of interest. Over the past decade, a growing body of research has emerged with the aim of addressing these issues. More recently, the MPEG community has started a new work item known as the "Multimedia Content Description Interface," or "MPEG-7" for short. MPEG-7 aims at specifying a standard set of descriptors and description schemes that can be used in describing various types of multimedia information ([58], [57]) to facilitate *Multimedia Information Retrieval* or *Multimedia Access*. The major objective of MPEG-7 is to make audio/visual material "as searchable as text." But since MPEG-7's main target is to define a multimedia content description interface, they leave other core tasks and content analysis techniques open, although the field has clearly defined a number of multimedia access tasks which have become the focus of recent research. Since video, which includes video, audio, and text, is an important and typical type of multimedia, we have made it our focus. Below we define and provide details on these different tasks as applied video data.

1.1.1 Video Database Indexing and Retrieval

Given a large collection of continuous video streams, the *video retrieval* task centers on retrieving relevant video in response to a user's query. Generally, systems built for addressing this task begin with *indexing* a collection of video streams according to the contents that the video contains. Such an index allows for quick identification of those media documents that satisfy the user's request.

In content-based video indexing and retrieval systems, video data are automatically archived and indexed by low-level features and thus we can search for a video in a large video database electronically based on the content analysis of video signals. However, much more flexibility can be provided for indexing and retrieval via a video database with video concept content classification. As a video's low-level features are generally high dimensional and difficult to describe in the query languages, the semantic classification of video can provide users with more friendly, convenient, and accurate responses to their queries. Conceptual content analysis for general videos is thus an active research area.

1.1.2 Video Summarization and Browsing

One of the important tasks in data compaction and access is to create summaries for video browsing. *Browsing* can be defined as the task of becoming familiar with the contents of a collection of documents in order to find individual or groups of documents that are relevant to a multimedia information need. An efficient summary can give an important clue about what is contained in the data without the necessity of viewing the entire session. Uchihashi et al. ([92]) developed a system called the "Video Manga" which generates semantically meaningful video summaries. For real-time collaboration sessions, video summarization can provide fast and efficient browsing and thus help users to catch up with past video content. For video databases, users can search information by browsing the summarized data; for example, the user would be supplied with a menu from which he could choose different video categories of interest which had been generated automatically by video semantic content analysis, thus saving search time and effort.

In order to facilitate video summarization and browsing, it has become necessary to provide infrastructure or computational tools that allow users to get quickly a better sense of the content of multimedia collections than is now possible. Even Yahoo!, which is a good example of current web documents browsing, provides manually constructed hierarchical topic directories. Thus, efficient and automatic video analysis tools which can help to organize and group video into categories in some semantically meaningful way are necessary to eliminate this laborious manual work.

1.1.3 Video Streaming and Filtering

It is getting simpler to produce, store, and transmit digitized video with the advent of multimedia-capable computers, cheap storage, and the proliferation of computer networks. Thus, streaming video is becoming an increasingly important type of data transmission on the Internet. Today, video over the Internet is used for different purposes, e.g., videophones, multiparty videoconferencing, transmission of seminars, conferences, live events, and so on. With the increasing prominence of the push technology over the Internet, streaming different types of video, such as news, will become common. The convergence of networking and TV is another important technological advance which requires efficient archiving and managing of real-time transmitted video data over the Internet.

In contrast to the "pull" applications of multimedia ([57]), many "push" applications such as user agent driven media selection and filtering, personalized television services and intelligent multimedia presentation follow a paradigm more akin to broadcasting, and thereby the emerging multicasting over the Internet. Such applications have very distinct requirements, generally dealing with streamed descriptions rather than static descriptions stored on databases, and thereby place more weight on selection and filtering rather than on indexing and retrieval. The "push" type applications consist of two primary collections of objects. First, many multimedia streams generated in real-time are to be disseminated to all users in various forms. Second, actual users differ in their specialized interests, degrees of interest, and abilities to participate in the multimedia stream sessions.

Different from the video query problem, in which a static collection of video documents is searched using dynamically generated user queries, the tasks of video *filtering* deal with a static query against which a dynamic incoming video stream is matched in order to capture new information relevant to a user's persistent information need. It is important to note here that the user's standing information need may be explicitly articulated as a query, or simply expressed via a *training-set* of video documents which may be categorized as binary decisions of either relevant or irrelevant by the users. In this regard, filtering is analogous to the task of classification or supervised machine learning, which will be discussed in later chapters.

In summary, content is the king in this information age. Effective content analysis for content indexing, management, and redistribution is the core technology for successful networking and information access systems. Information indexing and retrieval, browsing, and filtering are the key tasks to satisfy users information needs and are the major challenges of multimedia information access.

1.2 Review of Previous Work

Video data and their content are layered in nature as shown in Figure 1.1. Video content can be represented at both the perceptual and conceptual level. In its perceptual presentation, video programs can be parsed and represented by video sequences/clips, video scenes, video shots, key frames, objects, and other perception properties, such as color, motion, or acoustic features [see Figure 1.1(a) and (b)]. The conceptual presentation of video refers to the content that is specified based on semantic meanings or conceptual categories or stories. Perceptual features are the foundations of semantic meaning expression, that is, the latter is built upon the former [See Figure 1.1(b) and (c)]. Hence, multimedia content analysis should be done at both of these levels.

This hierarchical video representation corresponds to a general conceptual modeling of video, which can also guide us in on-line video content

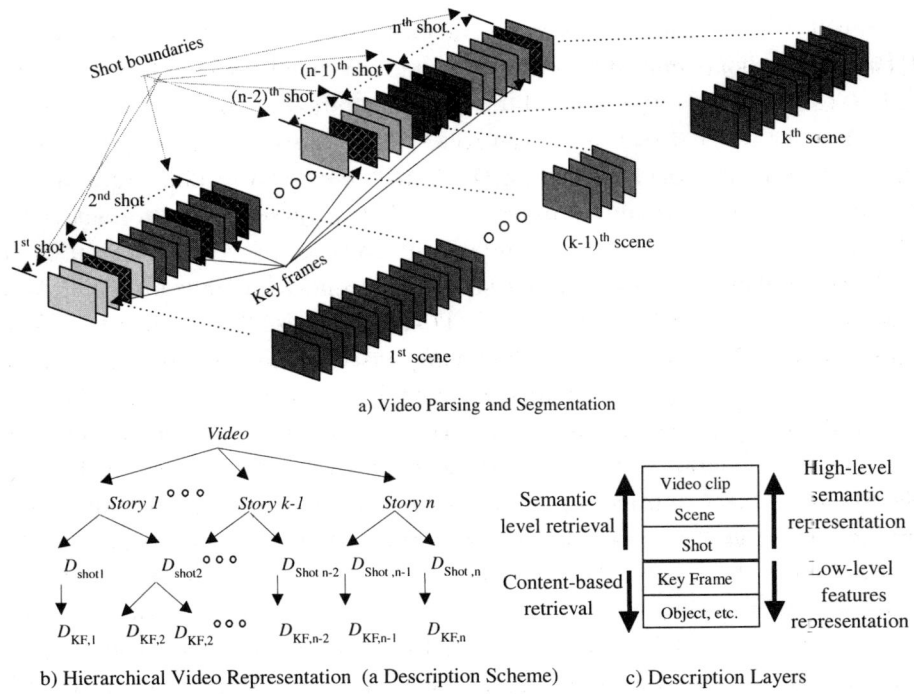

Figure 1.1. Video content parsing, representation, and description.

analysis. We see from Figure 1.1(c) that semantic meanings are generally associated with video shots, scenes, and clips, while key frames and visual/motional features are more closely associated with low-level representations which do not contain much high-level semantic meanings. A standard digital video consisting of a sequence of images doesn't contain any explicit information regarding individual units such as the different scenes and shots of which it is comprised. Ideally, for the purposes of data management, video data should contain additional information such as pointers which demarcate different shots as well as some attributes to convey the content of shots. Thus, video needs to be processed in order to become a representation suitable for searching/filtering and browsing. Video structuring methods have been described and discussed by several researchers ([30], [91], [86], [101]), and basically consist of the following two parts: physical parsing of the video to get low-level features, and conceptual modeling and analysis of the video to get high-level semantic meanings.

Many existing video database management systems ([86], [30]) support content-based queries based on low-level features. Specifically, Chang et al. ([12]) uses visual cues to facilitate video retrieval and Deng et al. ([22], [23]) described an object-based video representation to facilitate queries on the object. Video parsing and analysis techniques are also widely used for video summarization and browsing. Yeung and Yeo ([100], [99]) presented video posters to compactly present and fast-browse pictorial content. Rui et al. ([77]) explored the automatic extraction of video structures from both the physical shots and the semantic scenes and developed tools that can construct a table of contents (TOC) to assist user's access. Ferman and Tekalp ([29]) described a clustering-based framework to segment video sequence and generate a visual summary for video management.

Besides indexing and retrieval with low-level features, researchers ([24], [42], [43]) have also studied video classifications based on low-level features. Sports are very popular entertainment worldwide, and many researchers have worked on various sports video classifications. For example, Saur et al. ([81]) worked on automatic analysis and annotation of basketball video. They used heuristics of the basketball structure to guide the classification. Both Gong et al. ([32]) and Sudir et al. ([88]) used model-based classification methods. Picard ([72]) discussed in detail models of video and image libraries, which are mainly based on the classification results of digital images. These systems are applied to video databases. Other applications such as real-time Internet video streaming generally require fast and efficient content analysis and semantic classification. Due to the large variety of video data, any one specific data model may not be effective for generic purposes. A generic, fast, and efficient framework for video classification is highly desirable.

Many researchers have studied various other media besides video to help video indexing and retrieval. It is, however, more challenging to include mixed media cues for on-line distributed video streams. Patel and Sethi ([71]) studied audio classification and speaker identification to determine the presence of different actors in isolated shots in a movie database, thus facilitating the retrieval from a video database by speaker names. Chen et al. ([13]) discussed a pretty complete form of mixed media access to multimedia databases. The types of media considered in ([13]) include speech, images of text, and full-length text. The extracted information based on the mixed media cues is organized in a metadata structure. In some examples, they also extracted locations of keywords in speech and images, identifications of speakers, locations of emphasized regions in speech, and locations of topic boundaries in text.

Nakamura and Kanade ([62]) studied video content by making correspondences between image clues detected by image analysis and language clues detected by natural language analysis. They did experiments on CNN Headline News and video segments with important events such as a public speech, meeting, or visit, and achieved a fairly good performance. But since natural language processing needs a collection of lexicons for different videos, it is neither flexible nor easy for on-line cases. Liu et al. ([52]) used audio and video cues and the Hidden Markov Model to classify video among different types such as commercial video, basketball video, football video, news video, and weather video. Since this kind of video classification is pretty coarse, finer classification based on audio should also be studied.

In summary, the issue of extraction of useful information from a digital video is crucial. From our review of the previous work, we notice that a number of paradigms and approaches have been devised. These include the following:

- similarity search with one or several high-dimensional feature representations and some approximation evaluation techniques for similarity queries

- semi-structured data models and their respective query mechanisms

- approaches to clustering documents based on content and other novel IR techniques, such as classification on a certain level of video conceptual groups.

Some of these approaches have become more mature these days. However, query evaluation with only one approach is not always sufficient to cover all information needs. Instead, it looks promising to integrate different approaches to obtain better results. Nevertheless, there are a number of open problems remaining, in particular, we would like to point out the following.

- semantics of such complex queries and filtering, which contains multiple semantic meanings

- quality aspects, e.g., which combination of techniques yields better results under which circumstances?

- establishment of user-interfaces in formulating complex similarity queries and expressing relevance feedback

- resolution of system support, architectural, and implementation issues.

This book will specifically address some of the above emerging issues besides undertaking video content analysis.

1.3 Information Dissemination by Multicast

Different from unicast, multicast protocol allows the same data stream be sent to a group of receivers and thus minimizes link bandwidth consumption, sender and router processing delay, and large data delivery delay. As a result, multicasting is a better way of distributing common data to multiple end-users. A practical implementation of the IP-based multicasting is MBONE, which stands for the multicast backbone. MBone is layered on top of portions of the physical Internet to support routing of IP multicast packets since this function has not yet been integrated into many production routers.

MBONE is becoming more widely used these days. One of its applications is to distribute wideband real-time multimedia streams. Figure 1.2 shows a live streaming-video feed using multicasting to send files to multiple viewers at the same time. The camera shooting the live scene sends its video and audio feeds to a server, which then distributes the data as a single stream into the Internet. With multicasting, the data is maintained as a single stream until it gets close to the recipients, at which point it is copied and redistributed. This avoids some of the problems caused by the increased bandwidth consumption that accompanies growing streaming-video usage, such as in unicasting when a different data stream has to be sent from the server to each recipient. Multicasting also helps in data scaling.

1.3.1 Collaborative Semantic Multicast

IP Multicast is an effective means for data dissemination and sharing among large user communities. Unfortunately, current multicast collaboration media tools such as vic ([54]), vat[1] and wb[2] focus on effective and efficient codecs specially suited for Internet-based transmission, but are generally insensitive to the content semantics. Therefore, the concept of content-aware multicast should be introduced to add semantic-based support to many aspects of collaboration streams (video/audio/text) while exploiting multicast – as and when appropriate – as an information dissemination mechanism for information-sharing during distributed collaboration.

---

[1]http://www-nrg.ee.lbl.gov/vat/
[2]http://www-nrg.ee.lbl.gov/wb/

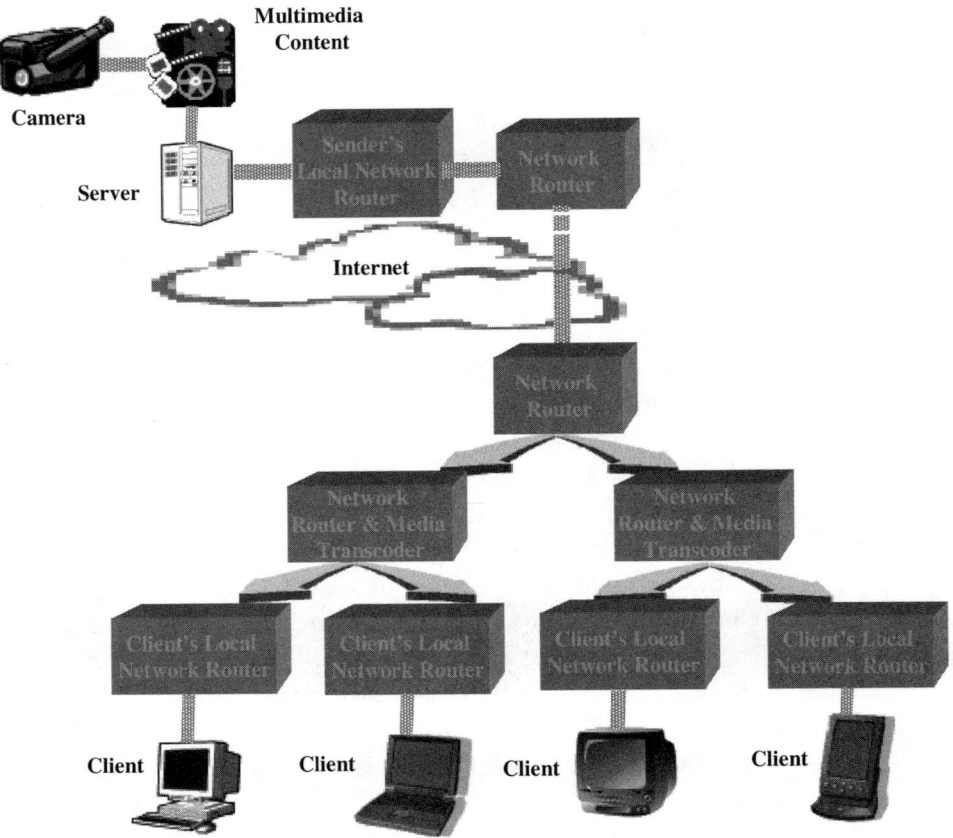

Figure 1.2. A live streaming-video uses multicasting.

Figure 1.3 shows a framework which requires a large-scale shared interaction infrastructure that provides a seamless environment for collecting, indexing, and disseminating the multimedia information produced in real-time collaborative sessions. This infrastructure captures the interactions between users (as video or audio streams) and promotes a philosophy of filtering, archiving, and correlating collaborative sessions in user- and context-sensitive subgroupings. Semantic multicast introduces a contextual focus along the multicast dissemination paths so as to reduce the information flow and to specialize the information to the needs and interests of particular groups of users.

The underlying basis behind creating a multicast service for collaborative sessions is to efficiently disseminate relevant information to every user

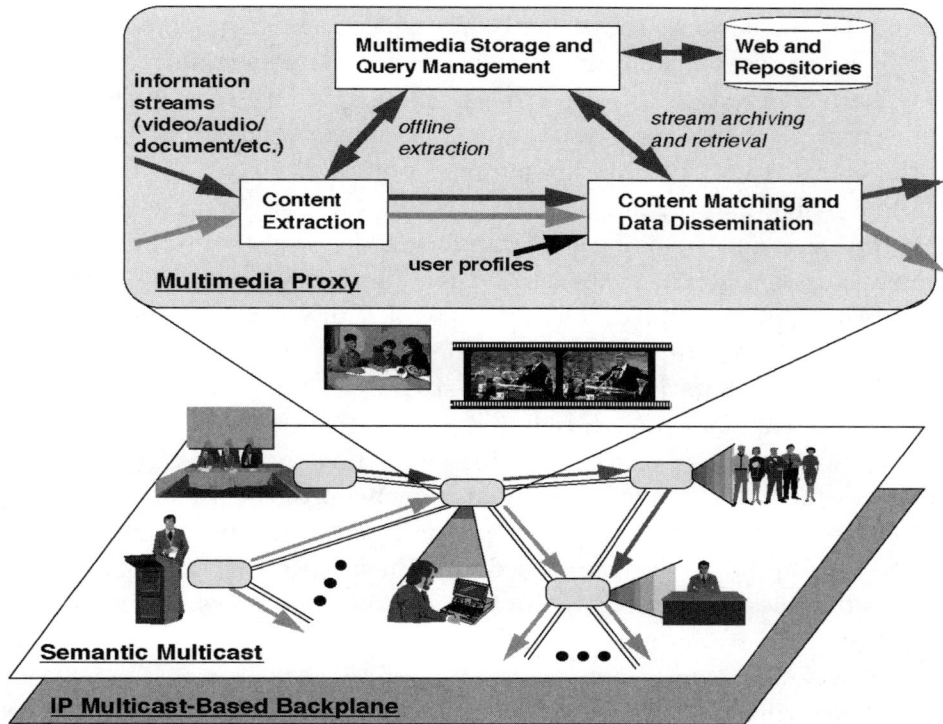

Figure 1.3. Information dissemination framework in content-aware (semantic) multicast.

engaged in the collaborative session. Current collaborative sessions usually consist of an "interaction stream" multicast over a single network channel. The scalable group communication model of IP multicast forms a natural basis for large-scale dissemination of information. Unfortunately, this network-level multicast protocol enforces a single level of indirection for delivering group data. In other words, if a user has an interest in the session, he must participate in the entire event and process all multicast information. Furthermore, if a collaborative session involves multiple interactions from interrelated working groups, a user must participate in the full broadcast from all groups to learn and understand the ongoing interrelationships. This model of collaboration only supports two modes of operation: either a user actively participates in the whole session or the user does not participate at all. In actuality, though, a collaborative session is one in which users participate in varying levels at varying times and multiple working groups, or interaction streams, concurrently overlap. Minimal support, if any, exists to decompose collaborative sessions among related working groups, filter and share information between groups, efficiently recall specific discussions, support occasionally disconnected users, or augment the interaction streams with "hooks" to on-line information repositories.

The goal of the semantic multicast system is to create a logical dissemination, filtering, and archiving structure for making the streams of the collaborative session available to the correct users in the right amount of detail. In other words, given a collaborative session of many overlapping streams, we automatically generate a semantic multicast graph which dictates the logical data-flow of collaborative streams, how streams should be annotated, manipulated, filtered, merged, archived, and propagated to effectively disseminate information to users belonging to different semantic interest groups. This contextual focus is accomplished by introducing proxy servers to gather and filter the streams appropriate for specific interest groups. Users are subscribed to appropriate proxies based on their profiles, and the collaborative session becomes a multilevel multicast of data from sources through proxies and to user interest groups.

To meet this requirement, every stream must be transmitted in real time or near real time, and quick content analysis and annotations are demanded so that the stream can be properly organized and disseminated in the semantic multicast architecture. The most important functionalities of content extraction, content analysis, and checking and data filtering as per user interests are carried out at proxy agents that are located in the networks. Here, the proxy agent is the generic concept of a machine which performs various operations needed for the on-line mode of data processing. The annotation proxy agent generates annotations, the filtering proxy agent passes or blocks streams based on whether or not they fit some scope, and the transcoding proxy (and/or the summarization proxy) transforms the raw data so as to support low bandwidth or logically disconnected users. It is possible to combine some of these operations into a single proxy. For example, the same proxy can perform annotations and filtering.

1.3.2 On-Line Multimedia Content Analysis and Annotations

Automatic Annotations are annotations which are generated by either processing directly on the raw data or on lower-level annotations. Some examples of these annotations are topic keywords summarization or key frame summarization of videos, video scene change tags to break video into visual meaning units, and concept identifiers of a given collaboration session, etc.

We distinguish between on-line and off-line video content analysis and annotations because the issues are considerably different in the two cases. Besides the challenges of off-line multimedia information management, there are unique characteristics of multicast video data that are relevant to management issues. For example, videos and audios are real-time continuous media in nature and present the difficulty of semantic interpretation in its raw form. Besides, streaming video is of a packetized nature, thus transmission over the Internet suffers from packet loss and there is no guarantee that all packets reach the receiver. It has to be kept in mind that data to be analyzed or stored in any multicast archiving system sometimes are incomplete and unconstructed as raw data. Following this section, we discuss some unique issues that are related to the generation of multimedia annotation in the on-line mode of operation.

Real-time constraint: First, it is imperative that the annotations be generated in real time since we need to perform some kind of filtering based on these annotations. This is especially important for data that are bandwidth-intensive, such as video, since there is a lot of data to process in very little time. The operations or algorithms should clearly be conductive to real-time processing of data. For video, operations such as processing in the compressed domain are desirable. Trade-offs of the complexity of the feature extraction algorithms and the effectiveness of the feature descriptors of semantic content analysis should also be carefully studied/evaluated for the on-line video content analysis.

Video semantic extraction constraint: Second, semantic extraction directly from visual data has traditionally been very difficult. Content-based (or feature-based) video retrieval is not efficient due to the lack of a comprehensive data model that captures structured abstractions and knowledge needed for video retrieval based on concepts. On the other hand, pixel-matching or feature-matching methods employed for query-by-example techniques are time-consuming, and have a limited practical use since little of the video object semantics is explicitly modeled.

The concept of layers of annotations: Third, not all annotations are generated by directly processing on the raw data. Typically, the annotations that are directly generated from the raw data, such as perceptual features, are very low-level and not directly capable of and sufficient for aiding high-level decisions. Hence, further processing has to be performed on the metadata or annotations to generate higher-level annotations such as the semantic categories of a video, which are better suited for aiding high-level decisions. For example, video key frames can be used for fast browsing, but key frames alone are still very inefficient for semantic interpretation of video as they are raw images too. Another example is the speech-to-text transcription, or text segmentation from video images, which is carried out directly on raw data to give a text transcript. To support higher-level semantic decisions, we might need more processing on the text transcript such as topic classification and summarization.

Uncertainty of metadata accuracies: Fourth, many of the operations which utilize metadata to do high-level processing tend to be very fuzzy or not very accurate. With the additional constraint of real-time processing, the metadata generated tend to contain a higher percentage of errors, which leads to the question: How low can the quality of the metadata be before it becomes completely useless? In this regard, since the metadata generated with the on-line mode is many times only used in a binary decision context as to whether the current data fit the scope or not, our filtering decision is relatively simpler than that of a retrieval problem and is unlike the digital library applications where metadata are used to select appropriate documents from a vast collection of documents.

Time granularity constraint: Fifth, annotations can pertain to different levels of time granularity. Some are indicative of the current time instance while others might cover a larger time interval, e.g., a text transcript in which annotations correspond to what was spoken "just now" as compared to topic summarization keywords or agenda which are indicative of more macroscopic time granularity. Thus, annotation format should be able to support a varying level of time granularity in video. There are many possible annotations to be generated. It will be convenient to have a format so that they can easily be inserted into stored data.

Annotation dependency constraints: Finally, many visual and motional features in video are based on multiple attributes or based on other feature-extraction operations, so agent cooperation and synchronization are necessary. Moreover, on-line videos normally cover a huge range of topics, which complicates the problem, so that we may need to learn what is of interest to the user from the user's profile, and then establish the knowledge base for that category of video. For example, if we know s/he is a CNN fan, we may establish a CNN news knowledge base off-line based on the characteristics of CNN news frames, such as the CNN logo, a model of the spatial structure of the anchorperson shots, the station background when the anchorperson is talking, etc. We may then use these characteristics to differentiate the CNN news video from other videos. So, in on-line video key frame and sequence classifications, knowledge memory and learning are very important. And, in general, how to establish the class basis and feature basis for a knowledge base from scratch within reasonable time duration is one of the most important research issues for on-line video classification.

1.4 System Overview

After discussing the challenges of multimedia information management in general and special issues of on-line multimedia content analysis and access over the Internet in detail, this book will address the solutions to some of those issues. We begin with an overview of a proposed layered video content analysis model and system for agent-based on-line video content analysis infrastructure that is applicable to both on-line and off-line video content analysis and access.

1.4.1 Layered Video Analysis Model

The "similar-to" operation is the most widely used in content-based video retrievals. It allows users to retrieve video and image instances that are close to the target (query-by-example) with a prespecified set of low-level features. However, it is still not easy or effective to retrieve multimedia data due to the lack of a comprehensive data model that captures the structured abstraction and knowledge that are needed for multimedia retrieval. To remedy such shortcomings of traditional content-based database management techniques and to satisfy on-line video analysis requirements, semantic inference or reasoning based on low-level features (e.g., looking for all video clips containing dunks, etc.) and conceptual meaning (such as fast action, slow movement, etc.) should be explored. The inference engine not only enables conceptual querying and filtering, but also allows automatic queries based on compositional features.

In this book, we present a novel knowledge-based system to support the Video Semantic Inference Engine as shown in Figure 1.5. It will bridge the gap between the video low-level features and high-level semantic concepts, which would facilitate semantic level query and filtering for on-line video dissemination. This book also describes the proposed a generic layered video analysis model as depicted in Figure 1.4. The layered video modelconsists of the Raw Data Layer; the Video Segmentation Layer; the Feature and

Content Layer; the Conceptual Model Layer, and the Knowledge Layer. Each layer is able to map to a module in the data dissemination model shown in Figure 1.5 and the data analysis flow chart in Figure 1.6. When we talk about a video analysis content model, we include audio and text information as part of the video sequence, and thus this model also applies to the audio and text analysis. The functions of each layer are detailed below.

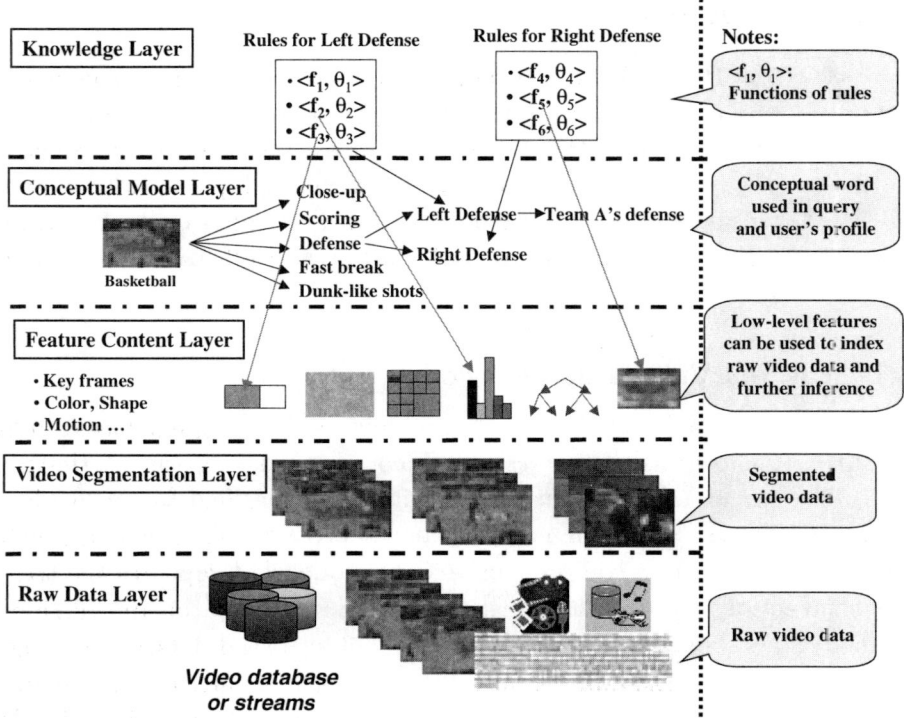

Figure 1.4. An example of a video layered model.

- *The Raw Data Layer.* This layer contains original video data which are either stored in the video database or received from on-line video streams such as video, audio, and captioned text streams from servers over the Internet. It is an abstraction of the coded video source in various formats that include audio, video, and captioned information. When a video is queried or matched with a user's profile, audio, video, and captioned data are multiplexed, and/or synchronized in transmission and presented to users.

- *The Video Segmentation Layer.* Different from the data in general relational databases, video is continuous media and is unconstructed. To understand any content of a video, or to analyze any video data, an efficient mechanism should be employed to parse the video into units which are either suitable for perceptual feature analysis or for conceptual abstraction analysis. For example, a video needs to be decomposed into physical units so that perceptual features, such as color extractions from key frames, will be analyzed while motion patterns need to be extracted from the video shots or scenes. As for the conceptual meanings of a video, we need to decompose the video into units which correspond to the conceptually meaningful abstract of the video content based on the conceptual model. Our system contains several levels of video content analysis proxies working on various levels of hierarchical video decomposition; they will be discussed in later chapters of this book.

- *The Feature and Content Layer.* This layer contains all the automatic video content annotation tags. Automatic annotations are those generated by processing either directly on the raw data or on lower-level annotations. Thus, the feature and content layer contains all the low-level video features extracted from the raw data, including physical video parsing tags and feature descriptors such as color, texture, and motion. The extraction methods and algorithms for the generic low-level features to be used for system evaluation will be described in Chapter 3. Correspondingly, our system contains several levels of video low-level analysis and extraction proxies working on various media, including video, audio, and text.

- *The Conceptual Model Layer.* Visual data in a video clip contain rich and unconstructed information. The Video Conceptual Model is an abstraction of the various visual data semantic types and their structures. In our system, the video conceptual model is mapped to a user's model, followed by the query engine and user's profile in Figure 1.5. For example, in the basketball video, most people may query certain key game events, such as scores and dunks. As shown in Figure 1.6, a hierarchy abstraction for each type of video for users' applications is proposed and presented. As a test-bed for our experiment and system evaluation, the conceptual model of the basketball video will be discussed in detail.

- *The Knowledge Layer.* The Knowledge Layer contains rules to map low-level features from each video clip to classes in the conceptual model. The knowledge rules are automatically learned from off-line learning algorithms in the off-line training module shown in Figure 1.6 and constructed as a tree structure. The feature attributes used for video classification are general and insensitive to the context. Visual entities in the Feature and Content Layer are linked to the Knowledge Tree in the Knowledge Layer to provide preset values for conceptual terms. This knowledge is used for conceptual inference in Figure 1.5 for further data dissemination decision-making. Also, the query engine can use the inference engine to automatically generate feature compositions for content-based retrieval with low-level features. High-level semantic content analysis agents, such as the video feature clustering agent and the video feature classification agent, analyze media semantics (such as the subject of a news video or the topic of the story of a video) exhibited by the stream.

1.4.2 On-Line Video Content Analysis Infrastructure

According to the layered video analysis model, a system prototype for a content-aware and user preference-oriented multimedia data distribution system on the Internet based on multicast protocol is proposed and described. Figure 1.5 shows an infrastructure prototype to support on-line media content analysis and present a model of service that realizes the goal of providing effective on-line content-based media dissemination by filtering. More specifically, we develop a real-time intelligent system prototype for fast video content analysis and dissemination over the Internet.

Collaborative sessions will generate various types of data streams, primarily raw audio, video, and graphics data, along with application-specific data types. As part of its operation, semantic multicast aims to filter, archive, fuse, and disseminate these data streams. However, many multimedia data streams in their raw forms are not amenable to automated semantic interpretation and typically have to be enhanced with other features, which are either manually created/attached or are extracted by analyzing the raw data.

This system will provide on-line feature extraction and semantic classification tools that will be used for data matching to the user's profile, plus filtering tools for real-time multimedia distribution and sharing over the Internet. This system is expected to provide synchronized multimedia

Section 1.4. System Overview 19

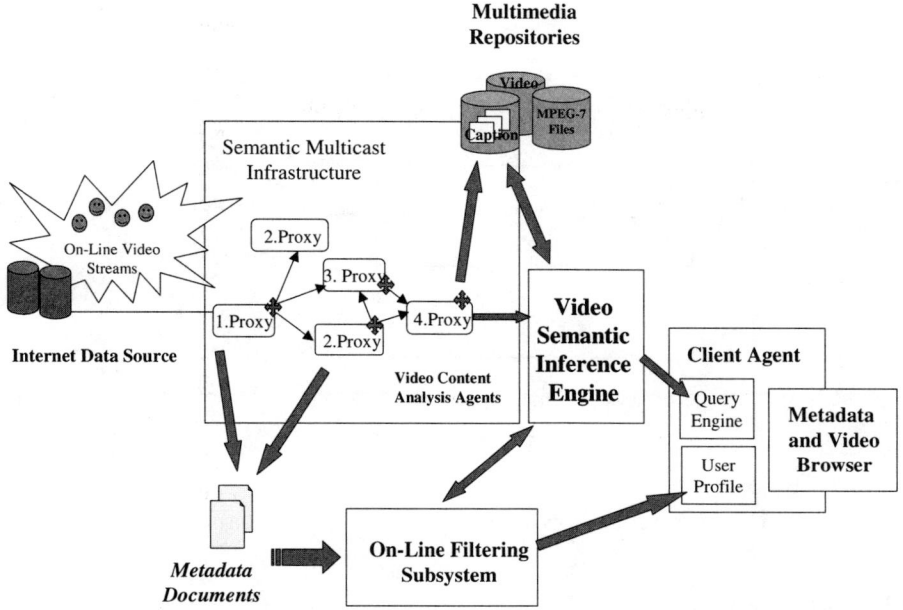

Figure 1.5. The on-line video stream analysis and dissemination infrastructure.

data stream distribution and filtering. In addition, the system will attempt to organize multimedia resources on the Internet in a scalable way, allowing users to find items related to their interest based on the content of the data. The system can also be extended to applications such as interactive and personalized TV broadcast services, training services, and collaboration applications.

Most functionalities required by content extraction, analysis and checking, and data filtering/dissemination per user's interest/query are fulfilled by intelligent content agents sitting over the Internet on the fly. Here, the proxy agent is the generic concept of a machine and software which are placed throughout the network and perform various operations needed for the on-line mode of operation. The annotation proxy agent generates annotations; the filtering proxy agent cuts off or passes through streams based on whether or not they fit some scope; and the transcoding or summarization proxies transform the raw data so as to support low bandwidth or logically disconnected users. It is possible to combine some of these operations into a single proxy. For example, both annotations and filtering can be performed by the same proxy. Furthermore, agents may archive information streams

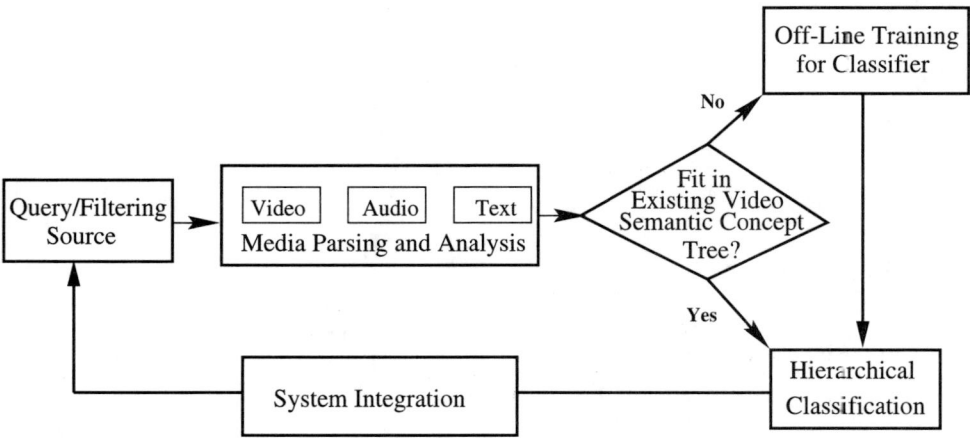

Figure 1.6. Flow chart of video content analysis.

and perform more detailed off-line analysis on the data to provide additional semantic structures for subsequent off-line retrieval.

The presented system makes use of formal machine learning, probability, and information theory to construct both perceptual and conceptual contents. Figure 1.6 presents a flow chart of video content analysis for users' applications. One key technical component is to decompose unconstructed video streams into structured units in both perceptual and conceptual meanings. While we discuss an integrated system more fully in Chapter 6 and show examples of the actual system usage, our presentation here provides a roadmap of technical components that comprise the system.

1.5 Roadmap of This Book

In this book, a novel on-line video content analysis system and a general layered video analysis model have been proposed and described. They provide guidance to the video analysis for management and access throughout the on-line video analysis procedure. We present in Figure 1.7 a roadmap of the technical components that comprise the on-line video content analysis system. Our research not only addresses the challenges arising from general video managements issues (either on-line or off-line) such as video semantic content analysis, but also stringent requirements imposed by on-line processing.

Figures 1.1 and 1.8 depict schematic analysis of the video structureand

Section 1.5. Roadmap of This Book

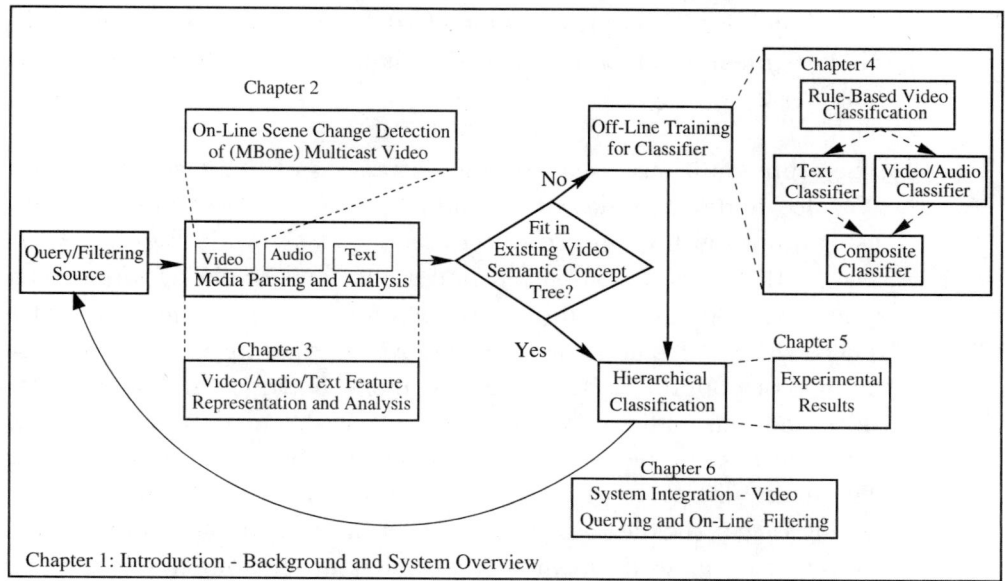

Figure 1.7. Roadmap of the book chapters.

the relationship between the video perceptual features and the conceptual contents. Video segmentation is followed by shot analysis in order to obtain the final structured video that contains link relations between different shots as well as content features for different shots. While perceptual features need to be captured from bottom up (Figure 1.8, Arrow 1), video conceptual meanings tend to be associated with larger blocks of video sequences and thus need to be analyzed from the top down (Figure 1.8, Arrow 2). With the layered video analysis model and the observations given above, hierarchical multilevel video segmentation and content analysis schemes have been implemented and presented in this book.

The video content analysis module is composed of three components as follows:

1. A video is analyzed so as to be segmented into shots. Shots can be defined as a set of contiguous frames which either depict the same scene, signify a single camera operation, or contain a distinct event or action like a significant presence or the persistence of an object ([30]). To initially segment a video, scene changes have to be detected and there are several schemes proposed to carry out this task in Motion

JPEG or MPEG compressed video ([103], [98], [83], [55]). Fast on-line video segmentation based on scene change detection is discussed in Chapter 2.

2. The representations, analysis, and extraction of the low-level features of video/audio/text data are studied. The low-level video features using a bottom-up approach are extracted as shown in Figure 1.8, Arrow 1. Here, a video is parsed into scenes and the key frames of the video scenes are extracted to represent and summarize the whole video. Then, the key frame images are analyzed by using image-processing methods to obtain features such as colors, textures, and shapes. Besides, motion, audio, and text features are also analyzed. These low-level features can be used to index video databases or link key frames. The main advantage of using low-level features in video database indexing is that the database organization can be completely automated. However, its main disadvantage is that it is very difficult to find the description of image/video, which is close to the semantic description of its contents. Nevertheless, conceptually similar video clips generally share common perceptual patterns, which provides the foundation of the video classification for conceptual meanings. This fact is exploited in Chapter 3 and Chapter 4.

3. High-level video meanings are analyzed by classifying video contents with a top-down approach (see Figure 1.8, Arrow 2) in Chapter 4. Since video may have a wide range of contents, effective classification must be done in a hierarchical way. The main advantage of the semantic description of a video is that it is possible to give very detailed and semantically precise descriptions of an image/video including terms which are unlikely to be determined by using video/image processing techniques alone. The main disadvantage of this technique is that, as yet, there are no fast efficient methods for video/image classification. To overcome this obstacle and to arrange a multimedia database into a more suitable organization for a human being's use, a rule-based classification system by supervised learning is described and a hierarchical video conceptual model specified in terms of a video concept tree is also discussed. Furthermore, novel machine-learning tools are developed to establish relations between the low-level perceptual features and the high-level conceptual video contents. The rule-based knowledge representation is novel and general enough to be used on a variety

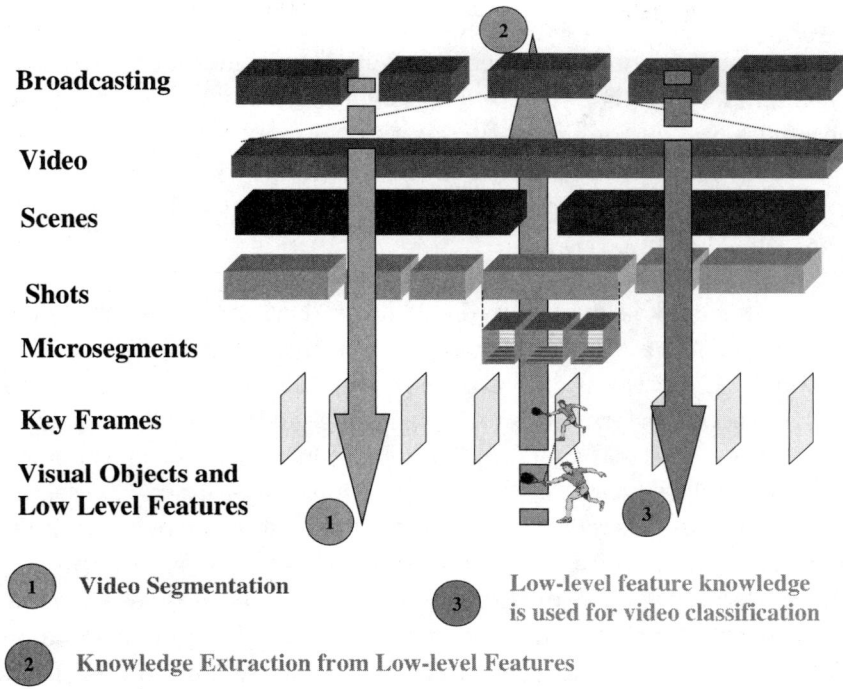

Figure 1.8. Video feature extraction and content analysis structure.

of problems in other application domains.

Having described all technical components in the system, we give the details of a simulation and its experimental results in Chapter 5. To demonstrate the application of the presented system, sample video clips of the different semantic categories are identified and appropriate low-level features are created. Off-line training is conducted to learn the knowledge and key features for each video category and thus a knowledge base is formed. Here, we utilized an entropy-based inductive tree-learning algorithm ([75]) as the learning algorithm. The knowledge base is represented as a decision tree with each node in the tree being an if-then rule as applied to a similarity metric utilizing an "appropriate" low-level feature along with a good "derived threshold."

We study and present the hierarchy decomposition of a CNN news and an NBC Olympic game broadcast, and then show how our approach recovers the structure automatically. Issues that are addressed include how to adaptively detect the anchor person and how to extract headline news stories by integrating both audio and text information. A table-of-contents of CNN news is generated to guide the user to specify the profiles for on-line filtering and off-line querying. For NBC Olympic game news hierarchy concept tree is generated based on the composite classifiers in video, audio, and text, and corresponding results are provided in this chapter, too.

1.6 Outline of This Book

The rest of this book is summarized as follows. Fast on-line scene change detection algorithms for video segmentation and indexing are studied in detail in Chapter 2. Experimental results show that the proposed algorithms not only provide good performances on scene change detections, but also demonstrate fast speed for real-time applications, which is novel and superior to previous work related to scene change detections for off-line video segmentations. After a video is segmented into different scenes, it can be summarized or indexed by the key frames extracted from each scene. Chapter 3 discusses the overall software and general algorithms that are used to generate the low-level perceptual features for content-based video indexing and querying. The video hierarchical concept tree is not only constructed from the video signal, but also constructed from audio and text. A novel tool system is also developed to fuse all this complementary information together to provide maximized information fusion. The construction of the hierarchical video classifier, text classifier, and fused composite classifier are discussed in detail in Chapter 4. In Chapter 5, some experimental results are conducted and reported based on the techniques described in earlier chapters. In Chapter 6, we give a detailed description of the whole integrated system in operation. We describe the overall semantic multicast system, and how our video content analysis infrastructure can be integrated into the whole system. In addition, two examples of applications are demonstrated for the integrated system. Finally, Chapter 7 provides brief concluding remarks on the general lessons we learned from the methods we proposed. We conclude this book by outlining additional potential applications and promising with directions for future research.

Chapter 2

ON-LINE SCENE CHANGE DETECTION OF (MBONE) MULTICAST VIDEO

2.1 Introduction

Internet packet (IP) video is emerging as an important multimedia application, especially with the increasing bandwidth available today. IP multicast is an effective means for data dissemination and sharing among a large user community. Over the past few years, real-time multimedia conferencing tools have been developed to operate over the Multicast Backbone (MBone). Current "multicast-aware" collaboration tools (e.g., vic ([54]) and vat[1]) only focus on effective and efficient codec but are unfortunately generally insensitive to content semantics.

Video is an important component of collaboration data streams. In order to enable filtering of on-line streams as well as the effective retrieval of desired data from an archive, the video/audio streams need to have annotations or tags generated that describe the "content" of the stream. In its raw form, multimedia data types such as video and audio are not amenable to automated semantic interpretation and typically have to be enhanced with additional information such as keywords. In addition, for summarization purposes and for further content extraction, the video can be tagged with scene change information as well as representative sample frames. There are some characteristics of video data for real-time content processing in proxies in the network. First video by nature tends to be relatively large in size as compared to other types of data. In addition, semantic extraction has traditionally been very difficult from visual data. On-line processing places

---
[1]http://www-nrg.ee.lbl.gov/vat/

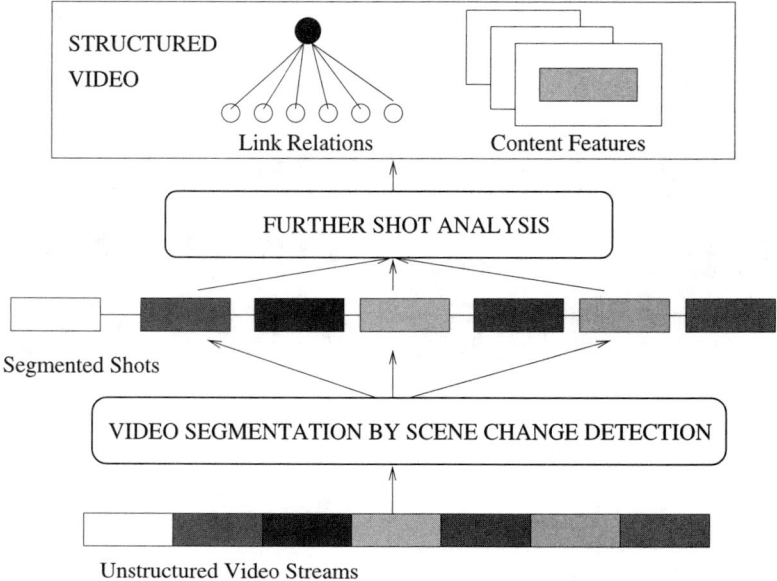

Figure 2.1. Video segmentation for video structuring.

very tight time constraints which affect the complexity of the algorithms. This is especially important for data that are bandwidth-intensive such as video. Operations or algorithms should clearly be conducive to real-time processing of the data. Note that content-based transcoding of the data can also be a filtering operation to assist users who have very low bandwidth. We distinguish between on-line and off-line annotations because the issues are considerably different in the two cases. One of the biggest differences is that off-line modes do not have the real-time constraint in generating the annotations. Thus, off-line annotations can be more sophisticated than those generated using on-line processing. Here we consider automatic on-line annotations on packet video. Automatic scene change description tags and key frames are examples of such descriptor annotations. To generate these tags, annotations based on raw video processing need to be utilized to detect whether a scene has changed and to generate a key frame that summarizes the scene. We evaluate different scene change detection algorithms with regard to different performance criteria on Intra-H.261 video.

Scene change detection is one of the most preliminary concepts in video structuring. There has been considerable work carried out involving scene change detection in both compressed and uncompressed video; however, there has not been any work done regarding analysis of video which has been packetized and distributed on-line.

Scene change detection is very closely connected to the compression algorithm involved. For example, several works have been published regarding MPEG compressed data. In this case, the data have been compressed using the Intra-H.261 scheme. Packets have been lost during transmission resulting in corrupt data.

Video can be multicast in different formats – H.261, Motion JPEG, nv, ivs, etc. However, currently almost 100 percent of the video telecast over the MBONE is based on the "intra" H.261 format. This is a special version of the H.261 code which does not use motion compensation. So, in this chapter, our on-line scene change detection is based on Intra-H.261 codec.

This chapter is organized as follows: Subsections 2.1.1, 2.1.2, and 2.1.3 provide a background on related protocols such as RTP(Real Time Protocol) and MBone related video tools. Section 2.2 describes related work on off-line scene change detection. The details of our system and the different scene change detection algorithms are described in Section 2.3. Section 2.4 gives the experimental results related to our different techniques. Section 2.5 concludes the chapter.

2.1.1 RTP Protocol

RTP ([82]), the Real Time Transport Protocol, has gained widespread acceptance as the transport protocol for audio and video on the Internet. It can be used for media-on-demand as well as interactive services such as Internet telephony and provides services such as time-stamping, sequence numbering, and payload identification. RTP consists of a data part and a control part called RTCP. The data part of RTP is a thin protocol providing support for applications with real-time properties such as continuous media (e.g., audio and video), including timing reconstruction, loss detection, security, and content identification. While UDP/IP was its initial target networking environment, efforts have been made to make RTP transport-independent so that it could be used, say, over CLNP, IPX, or other protocols.

Figure 2.2 shows the RTP header. The first twelve octets are present in every RTP packet, while the list of CSRC identifiers is present only when inserted by a mixer. Among all fields, marker (M) takes 1 bit. The interpretation of the marker is defined by a profile. It is intended to allow significant events such as frame boundaries to be marked in the packet stream. In the vic codec, the marker (M) is set on if the frame is a frame boundary.

Also, each RTP packet has a distinguished 16 bits sequence number. The sequence number increments by one for each RTP data packet sent, and may be used by the receiver to detect packet loss and to restore packet sequence. Another important field in the RTP header is timestamp, which has 32 bits. The timestamp reflects the sampling instant of the first octet in the RTP data packet. The resolution of the clock must be sufficient to achieve the desired synchronization accuracy and to measure packet arrival jitter. Generally, timestamps of the packets that belong to the same frame should be the same.

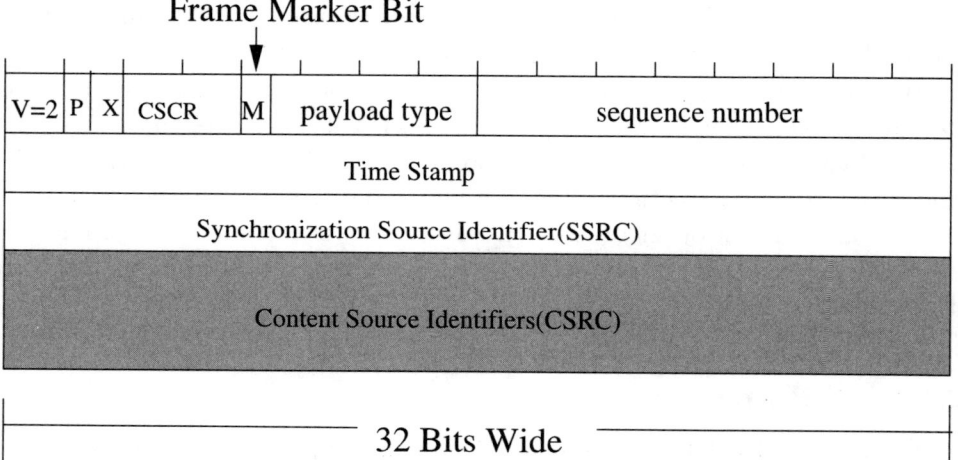

RTP: Packet Format

Figure 2.2. RTP header.

RTP is an unreliable protocol; all the packets generated at the source are not guaranteed to reach the receivers. The MBONE is restricted to 512 Mbps. Typically, there are packets lost due to congestion in the networks such as queuing losses at the routers, link losses, etc., especially for high bit rate data such as video. Thus, there is less than perfect reconstruction at the end points. Also, the rtptools ([82]), developed by Henning Schulzrinne et al., consist of simple command-line tools which allow the saving and playing of RTP sessions. There is indexing or interactivity support which will allow random access. Also, the tools have an intra- or intermedia synchronization function.

2.1.2 MBone Tools

The MBONE is growing day by day and soon it will be used for many different purposes. The basic MBONE tools can be characterized by a session directory application, videoconferencing and audioconferencing tool, and a whiteboard application. Currently, there are several IP multicast applications[2] which are being used for collaboration over the MBone – vic ([54]) for video, vat and rat for audio, mb and wb for whiteboard, and so on. Note that these are simply raw data streams which are encoded and transmitted at the source site and decoded at the client site. These raw data streams by themselves are not sufficient for making any kind of sophisticated decisions at the semantic level such as whether the current data fits a particular scope as defined in a user profile. We need to process these raw data to arrive at new representations which will enable semantic specific operations. We call these new representations metadata or annotations. The need for carrying these annotations after they are generated along with the raw data streams necessitates a channel or stream which accompanies the raw data streams of a collaboration session.

Furthermore, with the proliferation of networked video on the Internet, such as MBone multicast multimedia, there is a big demand to store/archive these videos for later playback and access. These multicast video-on-demand servers will need to support functionalities such as processing/analyzing the recorded data, creating indices, scheduling playback of data as per user requests, and so on. One of the standards which is gaining prominence is that based on RTP (real-time transfer protocol) and that based on IP multicast, which is an efficient method of delivering the same data to multiple users. There have been efforts published earlier which concern simple storage and playback of MBONE data. This can be distinguished into two parts – MBone recorder/players and archival systems, with the latter providing more functionalities than simple record and play.

MBone VCR ([37]): As the name stands, this application provides a VCR interface to control synchronized recording and playback of sessions sent over by the MBone. In this tool, the session is stored by synchronizing different media streams based on information provided by RTP. The MBone VCR provides some amount of indexing based on the timestamps so as to provide random access in addition to fast-forward and reverse capabilities. Some other features of the MBone VCR included blank-skip mechanism where parts of a session with no audio could be skipped automatically, as well as

---

[2] http://nic.merit.edu/net-research/mbone

automatic recording timeout, provided no data was received for a certain amount of time. Additional features included indexing, muting of individual media streams, and a primitive macro language. It also provides very limited indexing being able to jump to the next speaker, etc., by providing rudimentary indexing based on the audio part of the recorded session data. A command language enables limited programming features such as scheduling recordings and playbacks at a later point in time. The MBone VCR currently cannot be controlled remotely.

Following are two examples of the MBone archival systems: mMOD ([70]), which stands for the multicast Media-on-Demand system, is an on-demand system for the recording of different kinds of media that are being multicast on the MBone. Users can request the playback of a session using a web interface for starting and controlling of running sessions. Similar to MBone VCR, the VCR module offers some recording options such as source-host filtering, sender-timeout, appending to existing files, etc., as well as playback options such as receiver-timeout, marker-based playback, and limited time-based access. The mMOD is a completely distributed system with the VCR being able to be controlled using multicast messages addressed to a VCR using a unique identifier. However, the mMOD system still does not support remote interactive recording.

Interactive Multimedia Jukebox ([1]) is a web-based interface for requesting and scheduling of recorded MBone data. The requested sessions are played out on two different wide-area channels and one local channel. Sessions cannot be controlled in any manner once they are started. The server utilizes a scheduling algorithm whereby multiple overlapping requests are scheduled on a first-come first-served basis.

While most of the MBone video recording and archival systems record and re-access based on RTP-based session information only, none of them can provide content-based index and retrieval. Thus, it is our major research goal to explore the video content analysis and classification to facilitate the video indexing and archiving for off-line video data and filtering and redistribution for on-line video data.

2.1.3 Vic and Intra-H.261 Algorithm

Vic ([54]), which provides the video portion of a multimedia conference, is based on RTP. The most used compression scheme in vic is that based on "Intra-H.261" which is robust to package loss. Intra-H.261 is a small subset of H.261 in that: (a) only intramode frames in H.261 are adopted and

(b) no error prediction computation is involved, i.e., no inverse quantization and IDCT. Intra-H.261 provides improvements in compression gain and substantial improvement in run-time performances. Traditional compression algorithms are designed for constant bit rate channels, but Internet is a harsh environment for compressed video signals because it has relatively high packet loss rate, and traditional compression algorithms like MPEG and H.261 cannot do very well in this environment. The reason is that these algorithms use motion-compensating prediction to remove temporal redundancy. Although resynchronization can be obtained using intra-coded frames, the resynchronization interval is usually tens or hundreds of frames which makes the possibility of an error-free interval very low, making it virtually impossible for high quality video transmission. One solution to this problem is through "conditional replenishment" in which motion compensation is retained. In this method, each frame is partitioned into small blocks and information about only those blocks that change beyond some threshold is transmitted. Each frame is split into macroblocks with a size of 16x16. Updates are always intra-coded to avoid error propagation in prediction. To decide whether or not to encode and transmit a block, the conditional replenishment algorithm computes a distance between the reference block and the current block, by means of which each macroblock is compared with a corresponding block in the previous frame. As is the standard practice with common motion-compensation algorithms, vic runs conditional replenishment exclusively on the luminance component of the video. To avoid background noise and motion artifacts, an absolute sum of differences rather than a sum of absolute differences is used. If the block of reference pixels is $(r_1, r_2, ..., r_n)$, the block of new pixels is $(x_1, x_2, ..., x_n)$, and the threshold is T, then the new block is selected if

$$|\sum_{k=1}^{n}(r_k - x_k)| > T \qquad (2.1.1)$$

A background processing continuously refreshes all the blocks in the image to guarantee that lost blocks are eventually retransmitted. This solution works well for videoconferencing for three reasons: First, blocks that need to be transmitted usually contain the motion portion. And due to the spatial locality of the motion, they will be transmitted again even if a packet loss occurred previously. Second, the typical scene in a videoconference is a small motion portion with large static backgrounds, making the conditional replenishment a suitable choice. Third, the computational complexity for the encoder is reduced.

The above block selection algorithm will cause noticeable blocking artifacts. This is because of the fact that the block selection hypothesis may take hold and some blocks may no longer get replenished even though the block continues to change. Hence, the final block has a persistent error with respect to the final static state. The problem is solved by the conditional replenishment algorithm. When the selection algorithm ceases to send a given block, that block is aged and resent at some later time. This algorithm can be illustrated in Figure. 2.3. Each macroblock in the image has a separate finite state machine and a macroblock can be in various states. A macroblock is encoded and transmitted only when it is in the shaded states. Whenever the block selection algorithm detects motion in a block, the state machine transitions to the motion state (labeled M). If there is no motion, the block is aged to A1, A2, ..., etc., until it reaches the age threshold (state AT), when a block is sent and in turn enters the idle state(I). Vic fixes AT at 31. Besides a fill process, which is a background process running to continuously refresh all the blocks in the image to guarantee that lost blocks are eventually retransmitted, vic also selects some number of idle blocks in each frame and spontaneously transitions them to the background state(BG). On the next frame, it reverts to the IDLE state. Blocks are only transmitted in the M, AT, and BG states at low, medium, and high quality, respectively. The high bit is set to indicate that the block is in one of these states and should be sent.

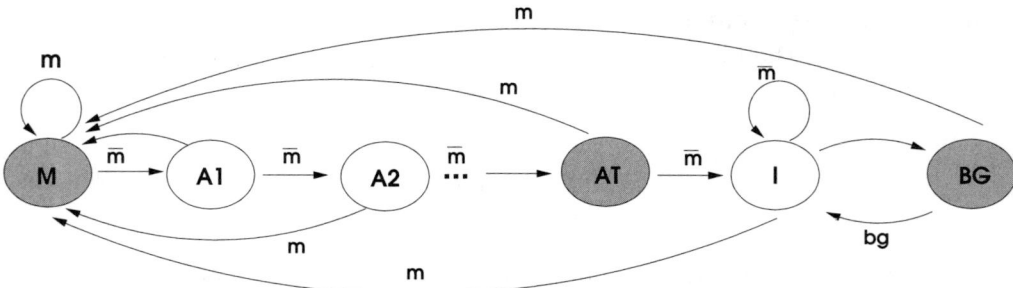

Figure 2.3. Block aging algorithm. A separate finite-state machine is maintained for each block in the image. State transitions are based on the presence (M) or absence (M-) of motion within the block. The block is replenished in the shaded states.

2.2 Related Work

The segmentation is normally accomplished by cut detection. Several approaches have been proposed, of which most fall into two classes: (1) those based on global representations like color histograms without any spatial information ([101], [59], [68]), and (2) those based on measuring differences between spatially registered features like intensity differences ([6], [4]). Apart from these approaches, shots may also be detected by finding changes in camera operation as well as on the basis of events such as the appearance/disappearance of an object, distinct changes in the motion of an object, etc. ([30]). Each of these techniques have their own merits, and while a new method of cut detection will not be a main component of our future research, we will compare some of the prominent techniques so as to choose an appropriate video segmentation algorithm.

Shot boundary detection typically represents the first step in structuring video sequences for content-based retrieval. Several techniques have been studied as discussed below for shot boundary detection, including pixel differences, statistical differences, histogram comparisons, edge differences, and so on.

There are several schemes based on pixel value comparisons between adjacent frames in a video. Zhang et al. ([101]) used a pixel and block comparison method on uncompressed bit streams. In this scheme, two frames are defined significantly different if the total number of the pixels of the two frames, whose pixel value difference is greater than a difference threshold, exceeds a certain percentage of the total number of pixels of the original frames. The disadvantages of this method is that it is not robust, is extremely slow, and manually adjusting the threshold is not practical. Shahraray ([84]) divided the frame into 12 nonoverlapping blocks and found the best matching from neighboring frames (same as in motion estimation). The weighted sum of each block difference provides the comparison measurement. Again, this method is quite slow and thus not suitable for on-line scene detection. Hampapur et al. ([34]) divided the change in gray level of each pixel between two images by the gray level of the pixels in the second image. This method is very sensitive to camera and object motion. Kasturi and Jain ([50]) used likelihood ratio comparison and computed the mean and standard deviation of the gray levels in different regions of the images. This method is better in noise tolerance, but is computationally expensive and also has a high false alarm rate.

Histogram comparison is the most commonly used method in shot detection. Histogram of frames are compared to detect an over-threshold difference. The difference of two histograms is calculated as the summation of all absolute difference between the two histograms on each gray or color level. Nagasaka and Tanaka [60] compared several simple statistics based on gray level and color histogram and found that the higher order of histogram comparison may not be more efficient than the gray and color histogram comparison techniques. The proposals of Zhang et al. ([101]) are considered here. All the algorithms are based on Intra-H.261 video over RTP protocol.

As mentioned previously, the Intra-H.261 algorithm only encodes the differences of macroblocks of the later frames from previous frames to get the most temporal compression rate. Three major goals that influenced our algorithm design are:

- On-line processing speed

- Accuracy rate and recall rate

- Packet buffering and synchronization

A major concern for our algorithm design is the scene change detection accuracy and recall rate . To save bandwidth for our on-line annotation, only key frames for further processing are sent out, so accurate scene change detection is very important as we don't want to waste our limited network and processing resources on the nonkey frames. Generally, there is a tradeoff between accuracy and recall rate. Our ultimate goal is to achieve the highest recall rate while still keeping the accuracy as high as possible. Recall rate is important for our on-line annotation and filtering proxies because the key frames are the fundamental elements for further annotation and filtering. In an on-line scenario, packets for video frames need buffering and synchronization. Intra-H.261 has the advantage over prediction-intensive codecs such as MPEG in that the packets of Intra-H.261 are all independent of each other, whereby the effect of any loss of packets will be limited to a few frames, and therefore has much better tolerance to packet losses than MPEG. Unfortunately, the compression and loss tolerance advantages of H.261 come at the cost of increased computational complexity. For the same reason, the decoding speed of Intra-H.261 should be much slower than MPEG codec too. So speed of video processing algorithms for Intra-H.261 is very essential to supporting real-time video summarization and annotation in on-line applications.

Data transmitted using a best-effort protocol such as IP suffers from packet loss. While transport-layer protocols such as TCP can handle this by asking the sender to retransmit the data, IP multicast typically employs UDP which does not guarantee delivery of data. Typically, in these systems, there are considerable packet losses due to effects such as queuing loss in routers, link losses, and so on. Thus, in our architecture, the data which are stored on the proxy is corrupt since they have packets which have been lost during transmission. During real-time playback, there is typically no time to take care of lost packets which results in poor playback performance.

2.3 Scene Change Detection Algorithms and Implementation

The basic framework of our proxies is the following: Our video scene detection proxy opens two sockets to the specific multicast addresses/ports. On the arrival of a packet, it is buffered for processing. Buffering is implemented using a linked list, ordered by the sequence number of each packet. Thus, packets which arrived out of order are taken care of during the buffering. The link-list buffer also keeps track of the frame that has just been processed. If a packet for that frame arrives after the frame has been processed, it will be discarded. In the meanwhile, to speed up the decoding and scene change detection, we keep packet buffering and packet processing in parallel by using multithreading programming in POSIX.1c.

Packet buffering includes packet copying/synchronization, while packet processing includes decompression/partial decompression, frame boundary detection/frame assembly, and scene change detection. When the packets are not always arriving in order, we must provide a mechanism to synchronize the frame without suffering too much on the processing latency. Packets are processed in an ordered manner from the buffer. We always check the Frame Marker Bit in the RTP header to see if a whole frame has arrived. If the last packet in that frame has arrived, we start with the scene change detection algorithm. In the meanwhile, we reset the timestamp for a new frame. If a frame has a few packets, then the first few packets of the frame generally have to stay at buffer longer than the last few packets. If a packet belongs to an old frame that has been processed already, that packet is discarded immediately. This explicit resynchronization of frame packets is due to the property of the Intra-H.261 framing protocol ([54]): packets in Intra-H.261 are independent of each other and can be decoded in isolation or in arbitrary order up to a frame boundary. To satisfy our real-time scene change detection requirement, we have implemented several scene change detection algorithms by combining off-line scene change detection algorithms ([101]) and coding features in the Intra-H.261 codec ([54]). Results of a detailed study and comparison of these algorithms are provided in Section 3.3.

The details of various scene detection algorithms we implemented are described below:

1. Full frame scene change detection (FDY and FDYUV)

    In this first implementation, we fully decompress every packet for the whole frame and only use the luminance values to get a 256 bin grayscale histogram over the entire frame. The dissimilarity of images measured by a luminance histogram is decided by Equation (2.3.1): given two histograms, H_i and H_j, each containing n bins, the normalized match index ([0, 1]) of the intersection of histograms is defined as:

    $$S(H_i, H_j) = \frac{\sum_{k=1}^{n} min(H_i, H_j)}{\sum_{k=1}^{n} H_i} \qquad (2.3.1)$$

    It is easy to see that $\sum_{k=1}^{n} H_i$ should be the total number of pixels in the frame. For our purpose, $d_c(H_i, H_j) = 1 - S(H_i, H_j)$ is the dissimilarity index to indicate dissimilarity between the two images. We call this algorithm FDY (full decompression with luminance histogram scene change detection algorithm).

    Besides using only a luminance histogram, we also implemented a color histogram scene change detection algorithm by taking both luminance and chrominance information into consideration for a color histogram. This variation of the first method is called FDYUV (luminance and chrominance histogram scene change detection method by using full-frame decompression). The FDYUV is almost the same as FDY. The only difference is that we consider color information by quantizing YUV into one byte with the bit ratio of Y:U:V =2:3:3, which are the 4, 8, 8 levels for Y, U, V, respectively. We found that this quantization of the YUV is adequate. In order to test the effect of chrominance information on the scene change detection, we put more weight on the requantization of YUV.

2. Frame Sampling (FSY and FSYUV)

    In this second type of implementation, we used frame sampling to speed up the processing procedure. Here, we called the algorithm which uses

Frame Sampling together with histogram comparison FSY if the histogram used in the histogram scene change detection algorithm contains only luminance information, and FSYUV if the color histogram contains all YUV information. We observed that shots with similar scenes would generally last at least two to three consecutive frames. So we chose the frame sampling frequencies of two frames/detection. The histogram scene change detection is the same as in the type 1 algorithm.

3. Partial Decompression Algorithms (PDY and PDYUV)

This algorithm is different from general scene change detection algorithms in the compressed domain because we take advantage of representational structure present in compressed video formats ([5], [103]). We are also constrained to standard streaming protocols for video on the Internet(RTP [82]). As mentioned earlier, Intra-H.261 only has an intra mode, so it doesn't contain any motion vector. Neither does it always send whole compressed data in a frame. But from the Intra-H.261 encoder's header, we are able to know the position and the total number of changed macroblocks. But this macroblock information is still too limited for us to judge whether it is a scene change detection or not, because macroblocks in a frame are sent when any of the following three reasons are satisfied. First, motion is caused by changes of macroblock content. Intra-H.261 only has an intra mode and there are no motion vectors for each macroblock. If there is a lot of object motion, most of the macroblocks in the frame are going to be changed locally and all of them must be encoded and sent out. Second, packet loss from the network and packet dropping from packet buffering will cause the number of changed macroblocks to not be accurate. Third, refreshments due to a conditional compensation algorithm make macroblocks received in that frame not always different from that of the previous frame. Intra-H.261 uses a conditional compensation algorithm to make up the quality losses caused by packet losses from network and aging blocks. From Figure 2.3 we can see that the conditional compensation algorithm is randomly refreshing those aging blocks and thus adds uncertainty to the macroblock numbers. From the above discussion, although we are not sure that all those received macroblocks are new with respect to the previous frame, we have observed that there is generally no scene change if only small portions (small number of macroblocks) of a frame are changed and encoded. So for those frames with few new macroblocks, we can save processing time by not doing full decompression and histogram scene changes, which are generally computationally expensive. Only for the frames with big numbers of changed macroblocks will we use either only luminance histogram scene change detection (here we call it PDY) or YUV histogram scene change detection(PDYUV) to do new scene detections.

4. Mixture method (MY and MYUV)

   Finally, we integrated a frame sampling algorithm, a partially decompressed algorithm and frame comparisons with a histogram scene change detection algorithm only as necessary to do scene change detection. We call this integrated method with Y histogram comparisons for those necessary frames MY, and the one with YUV histogram MYUV.

2.4 Experimental Results and Discussion

2.4.1 Results for Algorithms

In our experiments, we encoded a 10-minute clip from CNN Headline News at different bit rates ranging from 64 kbps to 1 Mbps. Here we used rtptools ([82])[3] to dump a 10-minute CNN news video with 1Mb/s, and then we used rtpgw tools ([2]) to transform the data into different bit rates from 64kb/s to 512kb/s. The processing was carried out on an Ultra-2 Unix Sun workstation with 256 MB RAM. Multicasting was done on our local network. We do not consider packet losses caused by the network. The Internet MBone may suffer from packet loss due to various reasons, with the primary reason being congestion in the routers. RTP (Real-Time Protocol), on which all the real-time media transfer in the MBONE is based upon, also does not provide any packet-delivery guarantees.

The performance was measured on the following four criteria: processing latency of packets, precision and output bit rate for the different bit rates and with the different algorithms, and the recall and precision of algorithms. The recall and the precision of algorithms are defined as:

- recall= correct detects/(correct detects + missed-detects)

- precision=correct detects/(correct detects + false-alarms)

---

[3]http://www.cs.columbia.edu/~hgs/rtp/

Section 2.4. Experimental Results and Discussion

Here, "correct detects" are the number of true scene changes detected by our algorithm, "missed-detects" are the scene changes in the video that are not detected out by our algorithm, and "false-alarms" are those video frames which are not new scene frames but detected as new scenes by the algorithm. The accuracy and recall rate measurements of different algorithms at different bandwidths are listed in Table 2.1. Table 2.2 shows results for the average scene change detection latency time of algorithms running with various bit rates (different bandwidths).

Table 2.1. Accuracy and Recall Rate of Scene Change Detection Algorithms (Bandwidth (BW) Unit: kb/s)

Algorithms	Accuracy for various BW				Recall Rate for various BW			
	128	256	512	1000	128	256	512	1000
FDYUV	97%	97%	97%	97%	90%	90%	90%	90%
FDY	90%	90%	90%	90%	98%	98%	98%	98%
FSYUV	96%	96%	97%	97%	76%	77%	80%	82%
FSY	83%	84%	86%	86%	87%	89%	90%	90%
PDYUV	98%	98%	98%	98%	87%	87%	87%	87%
PDY	86%	86%	86%	86%	95%	95%	95%	95%
MYUV	55%	58%	60%	60%	99%	98%	98%	98%
MY	50%	52%	55%	60%	99%	98%	98%	98%

Latency time is generally used to measure the speed of the algorithm for real-time applications. Here we defined the latency time of our scene change detection algorithms as the average buffering time of the last packet in each frame. For the last packet of every frame, we designate the time right after our processor receives that packet from the network as the starting buffering time of the packet, and we identify the time when our processor sends out the packet to the metadata channel as the stopping buffering time. The average buffering time is the total buffering time for the last packets of all frames averaged over the frames. One of the most important issues in the real-time processing algorithms is real-time video processing latency. In the video processing, reverse DCT and color requantization are generally the two most computationally expensive parts. This can be seen from our latency measurement in Table 2.2. The results also show that a color histogram generally takes longer than a luminance histogram comparison

Table 2.2. Time Latency (s) of Algorithms Versus Different Bandwidth (Bandwidth Unit: kb/s)

Algorithms	Accuracy for various Bandwidth (kb/s)				
	64	96	196	256	512
FDYUV	0.123	1.09	92.6	184.3	296.7
FDY	0.996	0.458	0.513	31.6	84.2
FSYUV	2.502	1.07	1.1	24.9	49.0
FSY	2.727	1.06	0.58	0.413	0.455
PDYUV	0.856	0.382	0.24	0.148	0.283
PDY	0.709	0.289	0.185	0.118	0.116
MYUV	1.16	0.513	0.325	0.219	0.211
MY	1.13	0.463	0.276	0.196	0.168

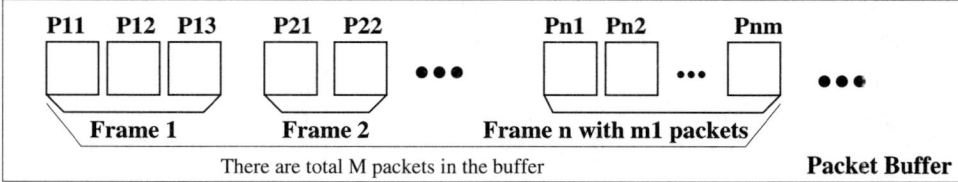

Figure 2.4. Packet buffering mechanism in our algorithm.

because we have to spend a relatively big chunk of the processing time on color requantization.

Figure 2.4 shows the packet buffering mechanism. Packets are buffered until each frame is finished with processing. The main processing for each frame is the total packet decompression time for all packets in that frame and the scene change detection time. Assume that there are $m1$ packets in the nth frame, with M being the total number of packets in the buffer. Also assume that the average decompression time for each packet is about TD and that the average scene change detection time for each frame is TS. Since the typical maximum packet size is limited to 1024 bytes, we can approximate the arriving rate r (packet/second) of packets based on the data bit rate and thus we know that about every $TR = 1/r$ second, the buffer will get one packet. Consider the following queuing conditions in the stored buffer:

- When packets are queuing up in the buffer, M is bigger than 0. Then,

Section 2.4. Experimental Results and Discussion

the buffering time for the mth packet in n^{th} frame P_{nm} is approximated as $BT_{pnm} = (M - m1 + (m - 1)) \times TD + n \times TS$. So within the same frame, the first few packets in the frame have to stay longer in the buffer than the last few packets.

- When the buffer always contains only one frame data, i.e., $M - m1 = 0$, the packet arrival rate is about the same as the packet processing rate. So the buffering time for the packet in frame n is given by: $BT_{pnm} = (m1 - (m - 1)) \times TD + TS$. Similarly, the buffering time for the m1th packet, i.e., the last packet, of the frame is: $BT_{pnm} = TD + TS$. Packet processing includes packet buffering/packet synchronization, decompression/partial decompression, frame boundary detection/frame composition, and scene change detection. The synchronization time of a frame generally requires that early packets have to wait for the last packet before the frame can be checked out.

- In the case where the bit rate is very low, often the buffer is in an incomplete state with respect to a full frame. Thus, the nth frame is not complete in the buffer, i.e., $M - m1 < 0$ and $m < m1$. In this case, the buffer time of packets in the frame is: $BT_{pnm} = (m1 - (m - 1)) \times TD + (m1 - m) \times (TR - TD) + TS$. However, the buffer time of the m1th packet(i.e., the last packet) of the frame is: $BT_{pnm} = TD + TS$.

From the above discussion, we can see that the buffer time of the last packet of each frame depends on the decompression time and the scene change detection speed, but does not depend on the bit rate. While the buffering of the packets other than the last packets depends on the all of the parameters of TD, TS, and TR, which cannot truly reflect the latency time caused by our video scene change detection processing. This is also very clearly shown in Figures 2.5 and 2.6. From Figure 2.5, we see that the latency time caused by the video processing should be the same for all last packets of the frames when video processing is fast enough. Typically, the early packets in a frame generally stay longer in the buffer than the last packet in the frame. When the bandwidth is too low and our processing is faster than the media flow, the earlier packets in a frame will have to wait for the later packet in the frame with the proxy processor in an idle state. We study the relationship of processing latency time and bandwidth (network bit rate) by measuring the first and last packet's buffering time with different bit rates. The results are shown in Figures 2.5 and 2.6.

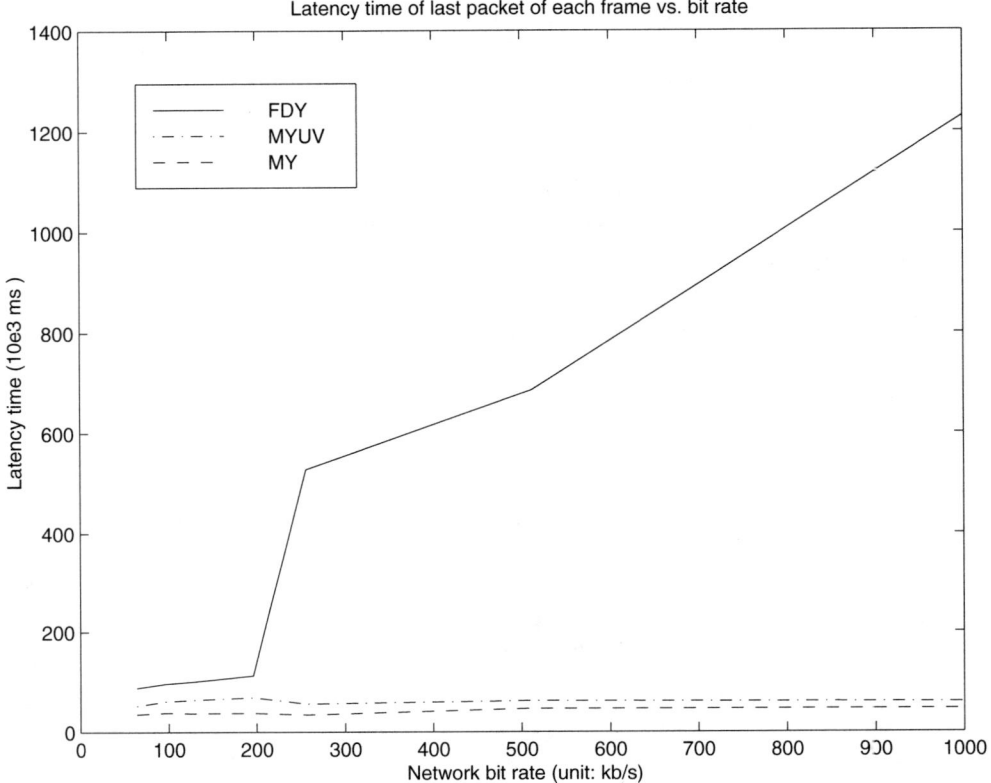

Figure 2.5. The bit rate affects processing latency.

We found out in our experiment that very low bit rate packet videos may not provide packets fast enough for our video processing so our processor has to wait for the packets to arrive. Table 2.2 and Figures 2.5 and 2.6 demonstrate this phenomena. Figure 2.5 shows the latent time of each frame's last packet with FDY, MYUV, and MY algorithms, and Figure 2.6 shows the latent time of each frame's first packet with FDY, MYUV, and MY algorithms. First, we noticed that both MYUV and MY processors are fast enough to keep pace with all bit rates from 64kb/s to 512kb/s video, so the latent time of the last packet in frames is flat. While the FDY processor works fine with a low bit rate, as bit rates go higher than 196kb/s, it is too slow to keep up with video packet flow queuing up the packets and causing the latent time of all packets going up. In an ideal situation where the video packets flow speed is the same as the processor speed, the latent time for the last packet of the frame should be flat, which is also true for the second to last packet of a frame whose latent time should be the same for all similar second to last packets, but has a longer latent time than the last frame packet. For MYUV and MY algorithms, it is in this ideal situation when

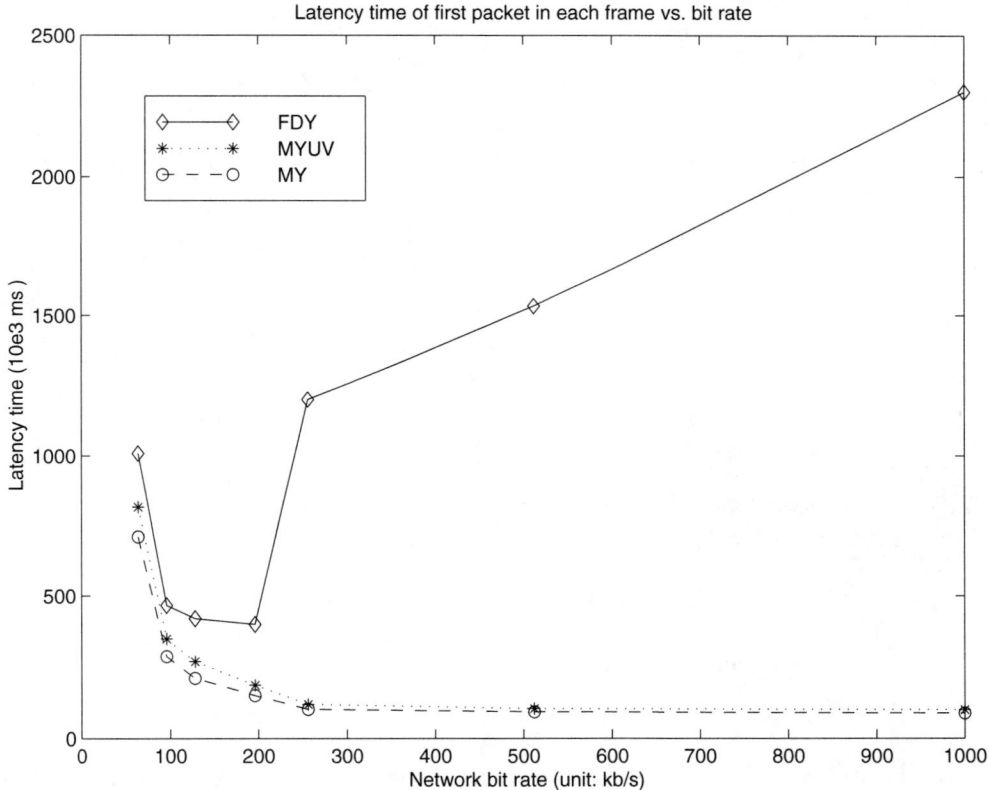

Figure 2.6. Bit rate affects processing latency.

data flow is higher than 256 kb/s. For FDY algorithms, the ideal data flow range is from 96kb/s to 200kb/s. After 200kb/s the processor of YDY is too slow to be real-time processing. Because the computer processor has to wait for the full frame packets before it can process on the frame, it generates longer latency time for very low bit rate videos due to the data flow itself which is caused by the rate control of the vic codec. From the measurement of the latent time of next packets from last packets in a frame, we can see that lower bit rate videos will make it wait longer for the last packet coming. From the above discussion, we can state that for lower bit rate videos, we

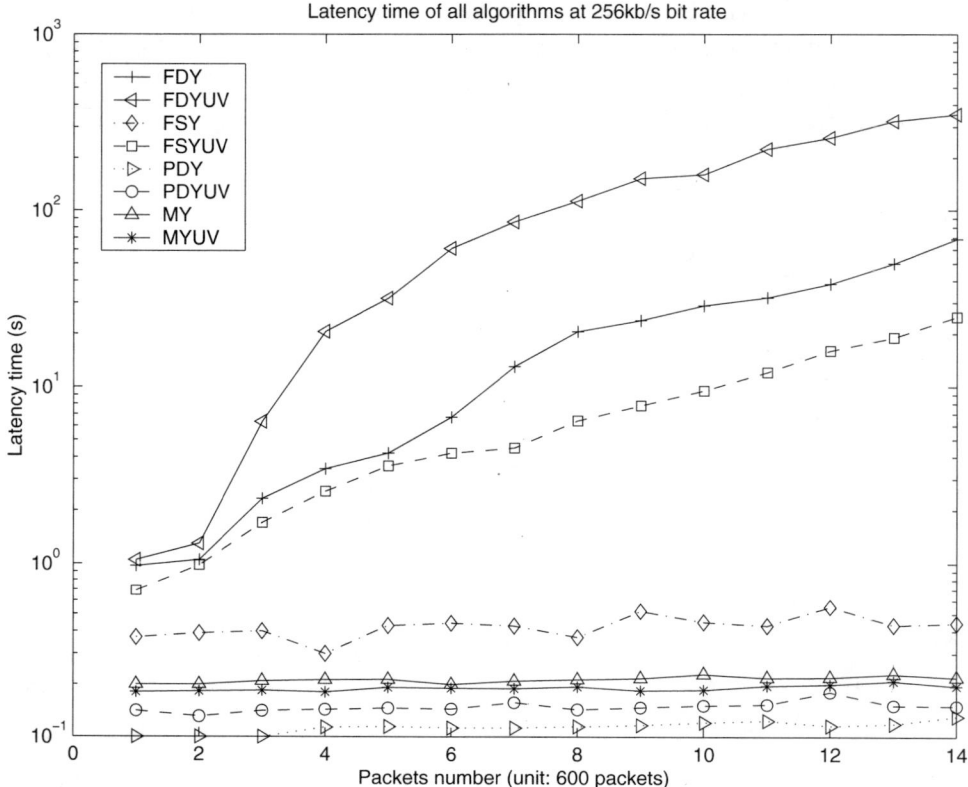

Figure 2.7. The latency of video processing.

have the option of using more complicated algorithms to get higher accuracy and recall rates. While for higher bit rate videos, the bottleneck is the processing speed, we saw that partially decompressed scene change detection algorithms, time sampling algorithms, and mixture algorithms will work fine with reasonably good accuracy and recall rates for on-line annotation and filtering.

Figure 2.7 shows the latency time versus the packet number. As we discussed before, if the scene change detection algorithm is fast enough to keep pace with the media flow, then every frame should have about the same processing latency. If this is not the case, then the later frames have to wait in the buffer and thus will have longer latency time than earlier frames. We also measured output data flow rate at different input data flow rate. The results are shown at Figure 2.8.

Section 2.4. Experimental Results and Discussion

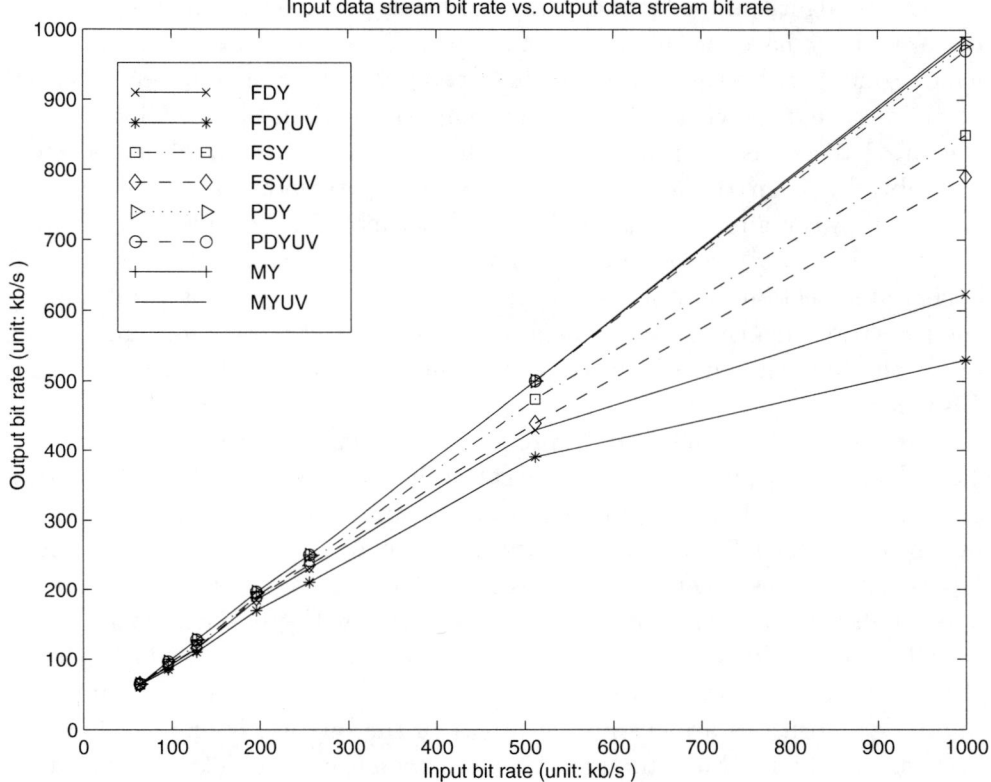

Figure 2.8. Output flow versus input flow.

2.4.2 Discussion

In this section, we are going to discuss the results we got from the experiment. The major issue is whether the algorithms meet with our design goals.

First, we compared accuracy and recall rates among all algorithms on four datasets with different bit rates. From Table 2.1, we observed that the histogram scene change detection algorithms are very consistent and the color histogram scene change detection algorithms are more accurate than luminance histogram scene change detection algorithms as expected. This is because algorithms with only luminance are not sensitive to color changes which in many cases represent object/background changes. Also, when only luminance intensity changes, such as "lighting," it would cause false alarms and thus decrease the accuracy. Among all the algorithms, those running on full decompressed frames have higher accuracy.

Among algorithms, MY and MYUV are the two with the worst accuracy; however, they have the highest recall rate. There are quite a few reasons which contribute to this: First, in these two algorithms, most scene change detections are done without histogram comparison. Frames with a lot of motion are detected as key frames because the algorithm senses a lot of changed macroblocks, even though these changes only correspond to macroblocks shifting between adjacent frames due to object motion. This problem is also caused in scenes which correspond to a dissolve object. As discussed in the background section, another reason for misdetection is that the conditional compensation algorithm will refresh aged macroblocks randomly and as a result this refreshment will increase the number of macroblocks and cause false alarms.

Frame-sampling can speed up processing and shorten processing latency. This algorithm has worse accuracy because the Intra-H.261 codec only sends out macroblocks that are different from the previous frame and thus any bypass of a video frame has the possibility of bypassing a scene change, especially in cases where the frames right after the bypassed frame have only small motion. The comparison of the latter with previous scenes may be below threshold. In this case, we will miss some key frames which affect the accuracy. But we also see that accuracy and recall rates for frame-sampling algorithms are getting better when the bit rate is higher. This is because at a higher bit rate video frames are more redundant and thus key frames are not easily missed.

On the other hand, the recall rate is a trade-off to algorithm accuracy. Generally luminance histogram scene change detection algorithms have a higher recall rate because they are using a looser criterion than the YUV histogram. FDY is the best algorithm when recall is important, while FDYUV is the best with accuracy and recall rate. But these two algorithms have the disadvantage of real-time scene change detection because they have much longer processing latency than other algorithms. Things get worse when the video is longer. After a certain point, the processing latency will exceed the tolerance.

The most important issue in real-time processing algorithms is real-time video processing latency. In video processing, generally reverse DCT and color requantization are the two most computationally expensive parts. The latency time measurement of frames is measured and shown in Figure 2.7. From the graph, we can see that the color histogram generally takes longer than luminance histogram comparison because we have to spend a relatively

Section 2.4. Experimental Results and Discussion

big chunk of the processing time on color requantization. Our experiments show that all of algorithms except FDYUV, FDY, and FSYUV are good for real-time processing with respect to the processing speed. They can keep pace with video flows ranging from 64kb/s to 512kb/s, which are the common bandwidth of packet video over the MBone. Figure 2.7 also shows that the latency time for all packets within the video using algorithms other than the above three are flat when the bit rate is at 512kb/s, which means the algorithms can keep pace with media flow. But this is not the case for FDYUV, FDY, and FSYUV. The graph shows that the later frames in a video data sequence have to wait longer to be processed because the processor is so slow that it causes the packets to queue up.

We also measured output data flow rate with different input data flow rate if the system is the real-time, then the output data flow rate should be the same as the input data flow, as shown in Figures 2.9 and 2.10. From Figure 2.8 we know that our fast algorithms (FSY, PDY, PDYUV, MY, and MYUV) can keep pace with input video streams up to 1Mb/s, while FSYUV, FDY, and FDYUV can only keep data flows up to 256kb/s.

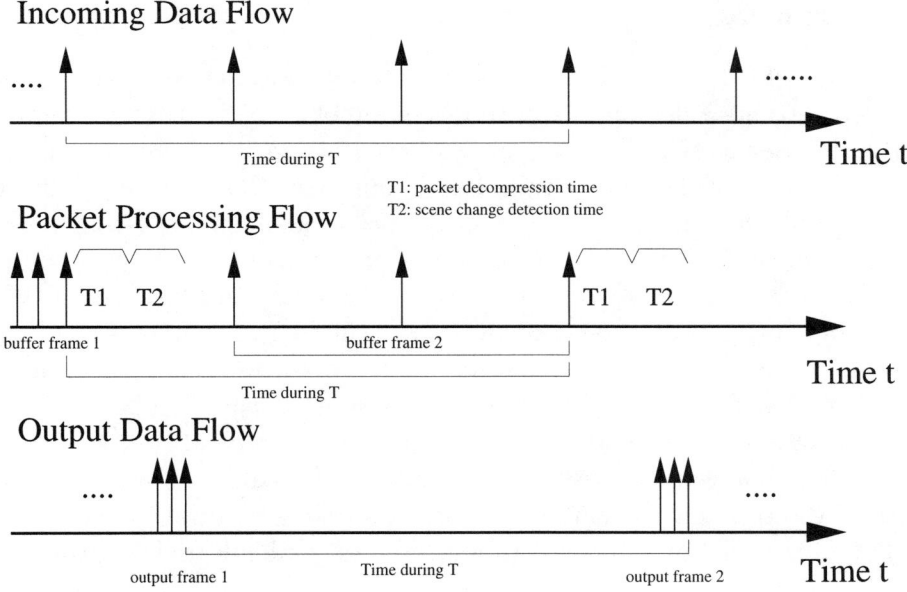

Figure 2.9. Data flow without video processing delay.

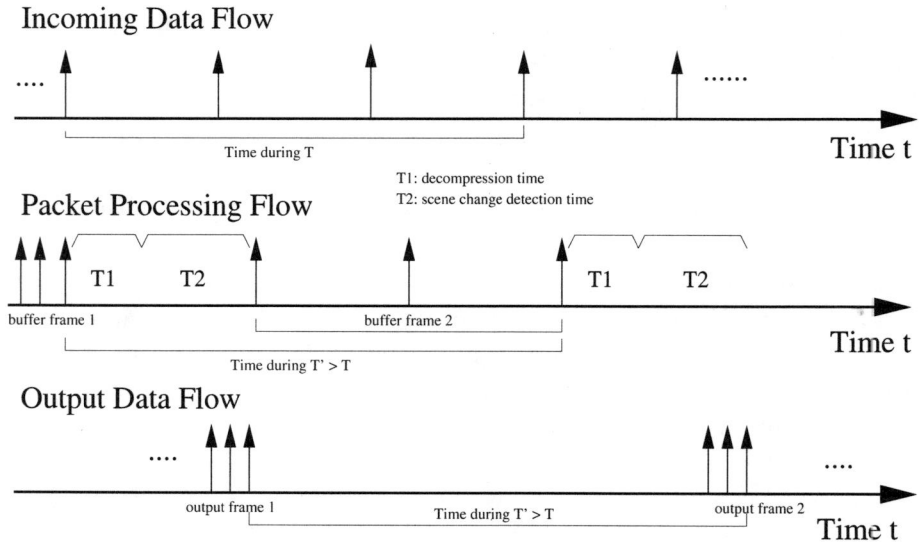

Figure 2.10. Data flow with video processing delay.

2.5 Summary

We presented a few effective methods of on-line video scene change detection over MBone video for the purposes of annotation/filtering. A detailed study of these algorithms with respect to accuracy, recall rates, and latency was given. One of the main advantages of our algorithms is their ability to support real-time video processing on the network. Joint algorithms based on video codec characteristics were carried out to acquire fast and accurate scene detection. Both global color histogram and block information of DCT coding were used. Performance of algorithms running with different bit rates was also studied. We presented results which proved that our algorithms are capable of satisfying on-line annotation needs. Also, with the detailed performance studies of these algorithms on different data bandwidths, we were able to choose a right algorithm for the best performance in each case. Because of the complicated nature of video processing, multiple workstations working in parallel may be used to realize more sophisticated real-time annotation.

In the next three chapters, we will study more video on-line annotation based on video content, such as video low-level feature extraction, key frame classification, and video sequence classification.

Chapter 3

VIDEO/AUDIO/TEXT FEATURE REPRESENTATION AND ANALYSIS

3.1 Introduction

Visual features are widely used and studied ([12], [23]) for video database indexing. Among them, the most popular features used for video database are color, texture, shape, segmented object, and motion features. Besides the indexing and retrieval based on the low-level features, researchers are also studying the video classifications based on the low-level features, such as camera motion types. But these classifications of low-level features can only extract a certain level of semantic meanings. Higher levels of semantic meanings contained in videos should be extracted in order to satisfy the requirement of on-line information dissemination and filtering. Another advantage of these low-level visual features is that they can be used as the feature sets of classification to extract high-level semantic meanings since similar story key frames and scenes generally share some of the same visual features.

Besides visual features, video is a rich multimodel source that also contains audio, image, and text information. Maybury, et al. ([53]) used mixed media cues to automatically segment video and analyze video content for summarization and visualization. Huang, et al. ([39]) used mixed cues of video, audio, and text to automatically segment news videos into different hierarchical levels, such as commercial breaks, news anchorperson, news summary, and news story sequences. So in our implementation of system prototype, we designed MPEG-7 compliable software not only to extract visual features, such as color, edges, and motion, but also to extract audio and text information to facilitate our intelligent rule-based video classification system which will be described in Chapter 4. All related works will be provided in each subsection.

3.1.1 Clustering in Feature Space

Once all the low-level content analysis agents generate feature values, clustering algorithms to group similar low-level features into groups are also designed. For example, for features of edges, we group edges into key frames with a lot of vertical edges, or key frames with a lot of horizontal edges. For key frames, we may also use statistical analysis to group the key frames into long lasting scenes or very short scenes. Long lasting scenes must contain more important information or slower motions than short scenes. Clustering can be used for summarization. There are several advantages and heuristic reasons why clustering is useful for visual, motional, acoustical, and text data. They are:

1. Clustering can be used to automatically group several images which are very similar to one another so that all the images belonging to a cluster can be treated as a whole unit. Similar video key frame images and motion patterns are expected to have features which are similar to one another. Thus, videos whose features cluster together in feature space can be expected to be similar and "clustered" or "grouped" into a category.

2. Clustering can be used as a method of organizing the feature space. Thus, a given feature space in a video can be treated as a subspace where each subspace can be a cluster or a cluster of clusters. This kind of partitioning of the feature space based on clustering can be a convenient abstraction for both users and the system. The cluster abstraction can be used to treat a large information space as a unit, without delving deeper into the content details of each individual image comprising the cluster.

3. Video/image/audio/text generally utilize features which are highly dimensional without adequate indexing structures. This results in an exhaustive search/matching of the total number of feature vectors. By using clustering and abstraction, the computational complexity of the search process can be drastically reduced. Clustering provides a means of organizing the feature space. Good organization should lead to better means of navigation.

4. Support semantically meaningful features. It can be expected that images which are semantically close to one another will cluster together in feature space so as to form "semantically meaningful" groups. These groupings can be based on different video/image/audio features such as overall color distribution, prominent colors, motion features, vertical and horizontal edges, etc. Thus, a combination of these features might provide a unique signature for a particular semantic group. We can summarize video in a cluster as being similar and can use some of the cluster characteristics to summarize/classify them. Again, clusters are a method of organizing the feature space.

5. Clustering can easily support browsing. In browsing, the emphasis is on quick perusal of the video. The human eye can easily summarize groups of visual data. In browsing, however, there should be very short latencies. By clustering together videos which are similar to a particular visual/motional feature, the user can easily browse in video.

Several studies have utilized clustering in general for carrying out content-based retrieval in images and video database. In ([106]), a top-down hierarchical clustering process, which adopts partition clustering recursively at each level of the hierarchy, is studied and used to build hierarchical views of video shots. The clustering process uses a fuzzy k-means algorithm at a higher-level of the partition tree and conventional k-means at the lower layers, where the clustering was carried out on the color histogram, temporal variance, and statistical motion features. The algorithm was used for video browsing and annotation.

3.2 Overall Software Structure

Figure 3.1 shows the basic procedures developed to segment and analyze video, audio, and text information. The Dali Library ([66]) is used due to its efficient abstraction and high performance routines in video, audio, and image data manipulation. First of all, video and audio signals are partitioned. Functions for video/audio random access, i.e., a frame-table or scene-table are built for MPEG-1 video for random selection/connections of video, and first developed as a library. Secondly, video shot detection is pursued to parse video into basic units, i.e., shot and scenes. A representative key frame is then selected for each scene for further visual feature analysis. In the meantime, audio signals are extracted and analyzed for special audio

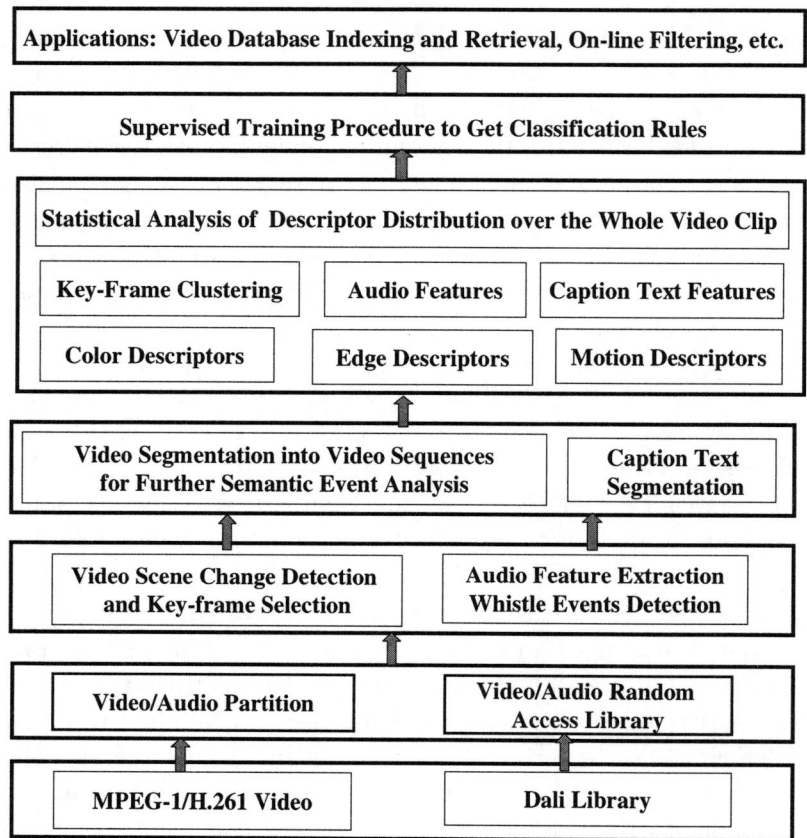

Figure 3.1. Architecture for video segmentation, analysis, and classification.

sound detection. In particular, we detect the whistle sound in a sports video for sports event segmentation. In addition, motion information is analyzed both at the key frame and at the clip level. Finally, caption texts for each segmented video clip are also segmented and analyzed.

In this chapter, a list of general features for video, audio, and text are described:

1. **Visual features.**

    - **Color descriptors.** This includes color histograms, dominant colors, spatial color, color clusters .

- **Edge and shape descriptors.** Edge features need clustering since it is very hard for a machine to describe any random edges due to high dimensionality and irregularity.

2. **Motional features.** This includes camera motion descriptor, motion activity descriptor, and object motion descriptor, etc.

3. **Acoustical features.** We discuss audio features in both time domain and frequency domain.

4. **Text features.** We discuss the text feature representation for text documents, which will guide us to text classifications for long text documents.

In the implementation of these video and image feature descriptor extractions, we concentrate on the simple and efficient extraction method and the descriptor design, but we don't define any matching metrics, which is important for video database retrievals. But as we are going to use the feature values to identify/infer video semantic categories, the more important issue here is the fast and accurate extraction of the values of these descriptors.

3.3 Visual Features Representation and Analysis

3.3.1 Color Descriptors

Color information plays an important role in object identification and recognition in the human visual system. Color information is also widely used in the video object segmentations on the key frame. The color histogram is one of the simplest and widely used features in video/image CBR (content-based retrieval) systems. But spatial information is generally lost in color histograms. For example, some visually different key frames might have exact color distribution, but they have totally different localized colors. So in order to get more detailed color information and increase the differentiation power of the feature, we consider the dominant color and localized color information too.

1. Color Histogram Descriptor

Color histograms represent the distribution of colors in images where each histogram bin represents a color in a suitable color space. Currently, the color spaces which are widely used in video coding and color information extractions include: R, G, B color space; Y, Cr, Cb color space; and H, S, V color space. All these color spaces can be transformed to each other. A pseudometric or pseudonorm, usually represented by a quadratic form, between a query image histogram and a data image histogram can be used to define the similarity metric between the two distributions. Since histograms are typically a high-dimensional distribution (e.g., N=256 colors), the distance metric is computationally expensive. Indexing on such high-dimensional features is typically not feasible, and in case of large databases, it is generally not feasible to compute the match measure against every image in the database. This is called the *dimensionality curse* in databases.

From the color histogram other color-related features can be derived, such as average color, color intersection, and color pairs. Color pairs were used in ([59]) to capture the spatial correlations between adjacent color regions. Extension of the color pairs to reduce the effect of background on the accuracy of color pairs was reported in ([15]). In [44], a similarity metric, which combines color information in terms of a histogram and shape information in terms of a histogram of edge directions is utilized to search a database of trademark images.

In ([102]), an extensive study using content representations based on color histograms is studied. Different color resolutions and restriction to dominant colors and matching based on both global and local histograms are studied. Suitable numerical index keys to support retrieval are also examined.

In one of the earlier works in this area, Swain and Ballard ([89]) first used color histograms as a feature to index images based on color. There are many advantages to using color histograms – histograms are easy to compute, are invariant to rotation and translation about the viewing axis, and change only slowly under change of angle of view, change in scale, and occlusion. They introduce the concept of histogram intersection as a metric to calculate similarity between histograms. They used the component color axes so that the intensity axis can be more coarsely sampled than the other two, because the intensity axis is more sensitive to lighting variation from shadows and distance from the light source. They claim good accuracy with 16x16x8 opponent color histograms. The computational complexity of this method is $O(N)$. They do not take into account the color similarity between adjacent bins of a histogram and use the L_1 distance metric.

In ([33]), Hafner et al. discussed the use of a low-dimensional, simple-to-compute distance measure between the color distributions to show that these are lower bounds on the histogram distance measure in fairly general cases. The similarity-measure based on the cheaper measure achieves both goals of low-dimensionality and completeness. Their similarity metric also takes into account the perceptual color similarity between adjacent bins of a histogram.

In ([31]), Gong et al. described a new method to create histograms which are compact in size and insensitive to minor illumination variations in highlight and shape. The histogram is created in the Munsell color space where the HVC color space is roughly subdivided into 11 groups where colors in each zone are perceptually similar. The advantage of this method is that it reduces the influence of noise and illumination changes on color histograms. They base their scheme on the observations that human beings tend to evaluate the similarity between images based on macro color configurations and image structures, while trivial color differences are not of much importance in the process. The other advantage of subdividing the color space into very few regions is that the resulting histogram dimensionality is much smaller and therefore easier to index. An indexing mechanism where each histogram is encoded into a numerical key and stored in a two-layered tree structure is introduced. The system is applied to a color image database containing around 500 color images. The downfall of this method is that it is prone to false positives or irrelevant images since the color space is subdivided into very few bins.

The color histogram descriptor is defined in Figure 3.2(a). In Color Histogram Descriptor, **HistogramNormFactor** indicates the possible value range of the **HistogramValue** values. **NumberHistogramBins** specifies the number of bins in the histogram. The number of bins is derived from the parameters of the Color Quantization Method it uses. **HistogramValue** contains the histogram values that are assumed normalized by the number of pixels to make the descriptor independent of the image size, and then further multiplied by **HistogramNormFactor**, which makes it possible to encode the histogram as an integer. The number of bits necessary to represent the histogram value is n, with condition $2n-1 < HistogramNormFactor < 2n$. For the color histogram descriptor, the extraction is a standard histogram extraction procedure and is shown in Figure 3.2(b). If histograms with different **HistogramNormFactor** are compared, they must be normalized by these factors.

```
ColorHistogramDescriptor {
    Enum    ColorSpace { rgb, yCrCb, HSV }
    Enum    ColorQuantization { linear, nonlinear, lookup-table }
    int     HistogramNormFactor
    int     NumberHistogramBins
    int     HistogramValue [NumberHistogramBins]
}
```

(a). Color Histogram Descriptor

```
numPixel = 0;
for (i=0; i<NumberHistogramBins; i++)   HistogramValue[i] = 0;
for (all pixels of a visual item) {
    i = QuantizedColorIndex(PixelValue);
    HistogramValue[i] ++;  numPixel ++;
}
for (i=0; i<NumberHistogramBins; i++)
    HistogramValue[i] = HistogramValue[i] x HistogramNormFactor / numPixel;
```

(b). Color Histogram Descriptor Extraction

Figure 3.2. Color histogram descriptor.

2. Dominant Color Descriptors

The dominant color descriptor specifies a set of dominant colors in any arbitrary shaped region. It is defined as Figure 3.3. In this descriptor, **DominantColorsNumber** contains the number of dominant colors in the region. **DominantColor** is a structure that holds the values of a single color and its percentage in the given region. The structure consists of an integer array **ColorValues** and an integer **Percentage**. **ConfidenceMeasure** describes the confidence measure on the calculated dominant colors. It can have values in the range of [0,100], where 100 means the highest confidence and zero (0) means no confidence, and minus one (−1) is used for cases where it is not computed (note that if it is not computed it does not mean that it is of low confidence). **ColorValue** is an integer array that holds the values of the dominant color in the selected color space. The dimension of this vector depends on the selected color space, which should be specified outside this

```
ColorDominantDescriptor {
    int     DominantColorsNumber
    struct  DominantColor   DCs[DominantColorsNumber}
    int     ConfidenceMeasure
}
struct DominantColor {
    int     ColorValue[ColorSpaceDimension]
    int     Percentage
}
```

Figure 3.3. Dominant color descriptor.

descriptor. **Percentage** describes the percentage of the region pixels that have that color. Note that the sum of the Percentage values for a given region does not have to add up to 100 percent.

The dominant color extraction method is as follows. If the key image has already been color quantized into a new image based on a given color clusters number, then the dominant color extraction is very simple and straightforward. It only has to calculate the histograms of each color cluster for the whole image, and the dominant color list is then ordered by percentage of the color. Let us assume that we have only color histogram $CH(I)$, denoting the percent of each color I in image object O. Let D be the distance in selected color space corresponding to "similar" colors. This distance will be used to threshold colors into similar ones that will then be averaged to produce each dominant color. This prevents averaging colors that are not close or similar. D is chosen to be constant for all colors, and is about 15 percent of the largest distance between any two colors in the color space. ColorDistance(C_1, C_2) is the distance between colors C_1 and C_2 computed in selected color space. The algorithm for finding dominant colors $DC(J)$ and their percentage $DCP(J)$ is as follows:

- Step 1: $J = 1$ (counter for dominant colors);

- Step 2: Get the most frequent color CHF from histogram $CH(I)$ where $CH(I) > T$; if none, END the algorithm;

- Step 3: Collect set of colors $SetC$ such that $ColorDistance(CHF,$

$CH(K)) < D$; for all K in color histogram $CH(I)$

- Step 4: Compute $DC(J)$ as average of color values of colors in $SetC$ weighted by their percentage value from $CH(I)$; Compute $DCP(J)$ as sum of $CH(I)$ values for all colors in $SetC$; $J = J+1$; Delete entries in $CH(I)$ corresponding to colors from $SetC$ (used to prevent counting colors more than once); Then go to Step 2.

T is recommended as five percent for, say, a 256 bin histogram. One can always keep T conservative and then take only dominant colors whose $DCP(J)$ is above a certain threshold. One way to compute the confidence measure is to sum the percentage of the dominant colors in the object. This will give high confidence to cases where pixels of dominant colors represent the majority of the pixels in the object, and will give low confidence to cases where there are spots of colors with small percentages, in which case most users might say there are no dominant/significant colors.

3. Regional Color and Corresponding Shape Descriptors

Zhong and Chang ([105]) applied color segmentation to separate images into homogeneous regions and tracked them along a time line for video content query. In their work, a uniform quantization in L:U:V color space was used. Kanai ([48]) used the uniform quantization in the HSV color space for the image segmentation. Both used the uniform color quantization to reduce the complexity of segmentation. Color segmentation generally can only segment out color regions instead of objects themselves because an object may contain a few different color regions.

In ([10]), the algorithm detects image clusters in some linear decision volumes of the L*a*b* space. The detected clusters are projected onto the line of the Fisher discriminated for 1-D threshold. The advantages of this method are that it permits the use of all property values of color clusters resulting in better segmentation, and is computationally efficient compared to other color clustering techniques because clusters are only specified in linear decision volumes using 1-D histograms.

The mean shift algorithm has been generalized by Cheng ([14]) for clustering data, and used by Comaniciu and Meer for color segmentation ([18]). The mean shift clustering method can be used to automatically detect a number of dominant colors and output them as a color tree for each frame. Here, a nonparametric gradient-based algorithm is used to provide a simple iterative method to determine the local density maximum. Then all pixels are backmapped into homogenous regions if the distance to the dominant color is not bigger than the threshold.

In our approach, the image is segmented based on the N color color-clustering and quantization done in the color clustering and dominant color extraction. Then the same color value pixels are backmapped into homogenous regions to get the regional color information and also the regions of the segmented area. The shape of the segmented color region will be analyzed for further information.

We also carry out edge detection on the clustered image. Note that by now the image has been considerably smoothed since it contains only a finite number of colors. By performing an edge detection on the clustered image, it is often easier to locate the boundaries by edge detection.

4. Clustering for Color Descriptors

Color clustering is useful for color quantization, color-based segmentation, and other color information extraction. There have been some earlier studies on using color-based clustering for content-based image retrieval. In ([49]), colors are clustered in the CIE uniform color space LUV, so as to get the few dominant colors in an image. A 3D RGB histogram of the image is computed followed by an analysis of the color peaks in the histogram, with each peak corresponding to a color cluster. Very small color clusters are merged together using a Euclidean distance metric in the LUV color space. Each pixel in the image is then assigned to the closest cluster. For each cluster, the mean and the number of pixels assigned to it is calculated. To reduce classification errors, the algorithm utilizes spatial information along with an optimal classifier which takes into account the population differences of the clusters in addition to the cluster mean. Each image feature consists of a set of clusters where each cluster is represented by the mean and the fraction of pixels assigned to that cluster. Two image feature cluster sets are compared using a method which takes into account both relative frequency of the pixels as well as the color distance between the two clusters. The algorithm was tested on a population of 170 trademark images with superior performance compared to histogram intersection and reference color method techniques. The computational complexity of this method is $O(m^2)$ for an $m \times m$ image and the cluster matching complexity is $O(pq)$ where p and q are the number of clusters, respectively, in two matched images. The computational complexity of this method is fairly high.

The color clustering technique for video key images in each scene is also used. The clustering is done in both LUV and RGB color spaces. The color clustering in LUV space is following ([49]) and the color clustering in RGB is using the media cut method in color quantization method ([35]). Currently, the procedure clusters each image into a given number of color clusters. The outputs are the images with final clustered colors. The clustered colors are both used in dominant color extraction and regional color information extraction.

3.3.2 Edge Descriptors

Changes or discontinuities in an image amplitude attribute such as luminance or tristimulus value are fundamentally important primitive characteristics of an image because they often provide an indication of the physical extent of objects within the image. Edges characterize object boundaries. They can be used for image segmentation, registration, identification, and representation of objects in scenes. Edge patterns can also be used to analyze key characteristics and scene classification. For example, in boxing sports, there are generally fence-like edges which contain more lines of horizontal edges. So edge patterns can also be treated as a simple texture in each key frame image.

The use of shape as a cue for indexing into pictorial databases has been traditionally based on global invariant statistics and deformable templates and local edge correlation. Sharvit ([85]) proposed an intermediate approach based on a characterization of the symmetry in edge maps. Sequential comparison of image visual features, so as to perform similar shape retrieval, is time-consuming and impractical. Access methods that utilize image shape features from different perspectives to narrow down the search space are necessary and essential. Wang et al. ([96]) described a two-stage matching scheme, combining global and local features, to enhance the efficiency of a search operation.

Our edge detection algorithm takes very common edge detection masks to fulfill the edge detection tasks. The edge detection masks include the first-order derivative masks, such as Prewitt and Robinson three-level masks, edge detection with an option of a second Laplacian mask is also provided. In the first-order derivative masks, all eight directional masks are used to detect edges along different directions. Once the edges of the whole images are detected, we then cluster the edge types into the horizontal edges, and vertical edges.

3.3.3 Shape Descriptors

As we are trying to extract and represent simple descriptors, here we will use the following parameters to describe a shape segmented by our regional color information. All shapes of the segmented region are represented as binary images.

1. **Area.** The most trivial shape parameter is the *area* of an object. In a digital binary image the area is given by the number of pixels that belong to the image.

2. **Normalized Area.** The *normalized area* represents the ratio of the area of the object to the whole size of the image.

3. **Center of Object.** X_m, Y_m. The center of the object is the average coordinates of all pixels of the object.

4. **Normalized Center of Object.** Normalize the center of object coordinates (X_m, Y_m) to the total image width and height, respectively.

5. **Moment-Based Shape Features.** For discrete binary images, the moment calculation is

$$m_{p,q} = \sum (x_1 - x_m)^p (y_1 - y_m)^q \qquad (3.3.1)$$

    The summations include all pixels belonging to the object. The moments with the zero-order moment, $m_{0,0}$, can be normalized to gain scale-invariant moments. Since the zero-order moment of a binary object gives the area of the object, the normalized moments are scaled by the area of the object. Shape analysis starts with the second-order moments. The first-order central moments are, by definition, zero.

6. **Object Orientation.** The *orientation* of an object is defined as the angle between the x axis and the axis around which the object can rotate with minimum inertia. The object is most elongated in this direction. Thus, the orientation of the object can be obtained by computing the eigen vector of the tensor of inertia of the object as given below:

$$\phi = \frac{1}{2} \arctan \frac{2m_{1,1}}{m_{2,0} - m_{0,2}} \qquad (3.3.2)$$

7. **Eccentricity.** The *eccentricity* ε is defined in Equation 3.3.3.

$$\phi = \frac{(m_{2,0} - m_{0,2})^2 + 4m_{1,1}^2}{(m_{2,0} + m_{0,2})^2} \qquad (3.3.3)$$

The eccentricity ranges from 0 to 1. It is zero for a circular object and one for a line-shaped object. Shape description by second-order moments essentially models the object as an *ellipses*.

Matching/Query process. With the above parameters of shape descriptors, we can match the similar images with following low-level queries served by the shape descriptor:

1. Find the 2-D visual objects whose orientation is similar to the object in a different image, either equal to a certain value or greater/lesser than a certain value, or are in a certain range.

2. Find the 2-D objects whose sizes (physical or measured as a percentage of the picture size) are similar to the one of this object, or either equal to a certain value, or are greater/lesser than a certain value, or are in a certain range.

3. Find the visual objects that are positioned near (x,y) (a particular coordinate) the location in the picture.

Once the descriptor values are available for each visual object, it is quite trivial to perform these queries. The descriptor Matching process is done by calculating the similarity distance. The similarity distance between two shapes is measured by summing the weighted absolute difference between two sets of feature vectors from two comparing images.

3.4 Motional Features Representation and Analysis

This subsection lists several descriptors that describe visual features related to motion. The motion descriptors here include the following: Camera motion/motion flow cluster descriptor, object motion descriptor, object motion activities, and motion flow cluster descriptor. There are other motion related descriptors, such as parameter motion descriptor, parameter trajectory (extension of parametric object motion descriptor), and object trajectory descriptor, etc., which are not going to be discussed due to the limitation of their usage for on-line applications because of high complication of the extraction of the descriptors.

Basic Camera Operation

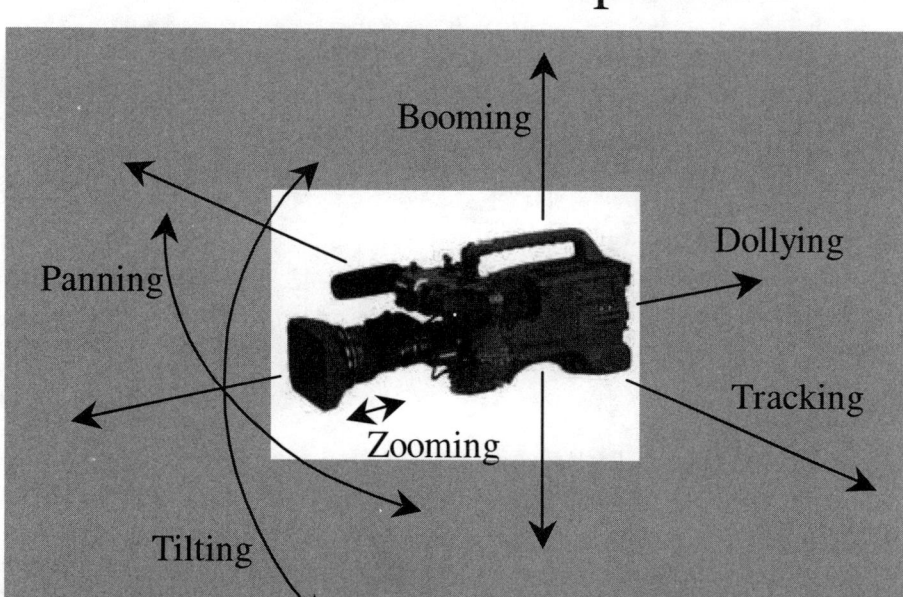

Figure 3.4. Camera operations.

3.4.1 Camera Motion/Motion Flow Clustering Descriptor

This Descriptor describes 3-D camera motion parameters. It is based on 3-D camera motion parameter information, which can be automatically extracted or generated by capture devices.

Camera Operations. The Descriptor supports the eight well-known basic camera operations (see Figures 3.4 and 3.5): fixed panning (horizontal rotation), tracking (horizontal transverse movement, also called traveling in the film field), tilting (vertical rotation), booming (vertical transverse movement), zooming (change of the focal length), dollying (translation along the optical axis), and rolling (rotation around the optical axis).

The subshots for which all frames are characterized by a particular type of camera motion, which can be single or mixed, determine the building blocks for this camera motion descriptor shown in Figure 3.6. Each building block is described by its start time, the duration, the speed of the induced image motion, by the fraction of time of its duration compared with

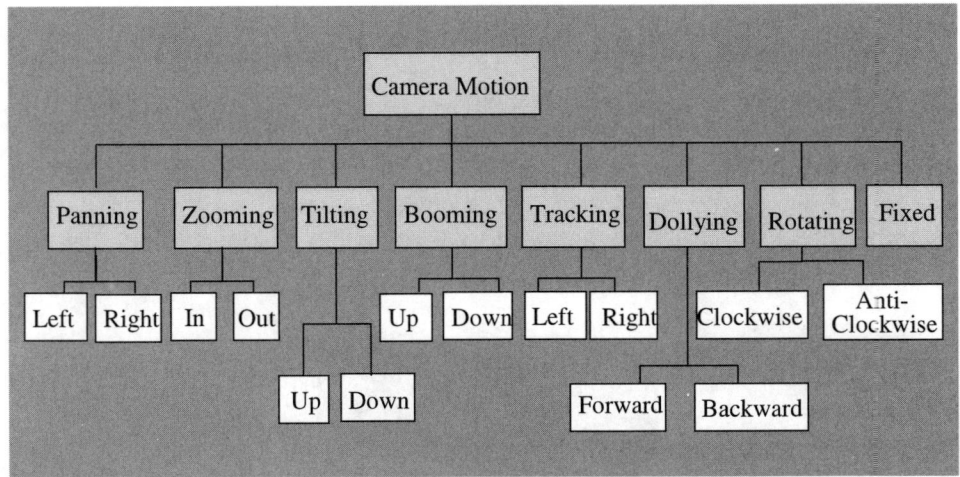

Figure 3.5. Basic camera motion catergories.

a given temporal window size, and the Focus-of-expansion (FOE) (focus-of-contraction (FOC)). The Descriptor (See Figure 3.6) represents the union of these building blocks, and it has the option of describing the mixture or nonmixture of different camera motion types. The mixture mode captures the global information about the camera motion parameters, disregarding detailed temporal information, by jointly describing multiple motion types, even if these motion types occur simultaneously. On the other hand, the nonmixture mode captures the notion of pure motion type and their union within certain time intervals. The situations where multiple motion types occur simultaneously are described as a union of the description of pure motion types. In this mode of description, the time window of a particular elementary segment can overlap with the time window of another elementary segment. The detail for these parameter extractions is following the camera motion flow model described in the papers ([87], [36], [45]). Some examples of motion flows for camera motion in frames are shown in Figure 3.7.

3.4.2 Object Motion Descriptor

A video sequence consists of several background and foreground video objects. Precise description of the motion in a video sequence is possible by describing the motion of the background and all foreground objects. The background motion is generally related to that of the camera. Description

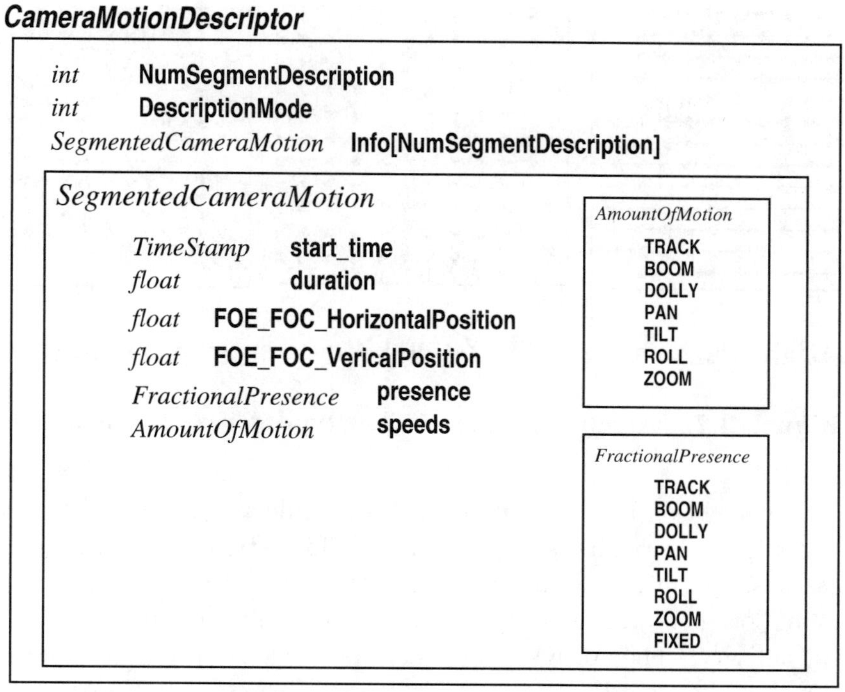

CameraMotionDescriptor

```
int      NumSegmentDescription
int      DescriptionMode
SegmentedCameraMotion   Info[NumSegmentDescription]
```

SegmentedCameraMotion

TimeStamp	**start_time**	
float	**duration**	
float	**FOE_FOC_HorizontalPosition**	
float	**FOE_FOC_VericalPosition**	
FractionalPresence	**presence**	
AmountOfMotion	**speeds**	

AmountOfMotion
TRACK
BOOM
DOLLY
PAN
TILT
ROLL
ZOOM

FractionalPresence
TRACK
BOOM
DOLLY
PAN
TILT
ROLL
ZOOM
FIXED

Figure 3.6. Camera motion descriptor.

of motion of foreground objects are generally more complex. Motion information of both background and foreground are very important for video content analysis. As Vic tools take Intra-H.261 codec, which lacks of motion vector for both camera motion/editing type and object motion information, we managed to describe the object motions by using an approximate measurement by considering the codec characteristics of Intra-H.261. As we stated in the scene change detection agent, we have changed macroblock information by simply calculating the center of changed macroblocks. We use the translation of the consecutive frames as object motion features.

3.4.3 Motion Activities Descriptor

This descriptor is compliant with the ongoing MPEG-7 developing motion activities descriptor ([25]). The essential attributes of a motion activity descriptor include:

1. *Intensity of Activity.* A high value of intensity indicates high activity while a low value of intensity indicates low activity. For example, a

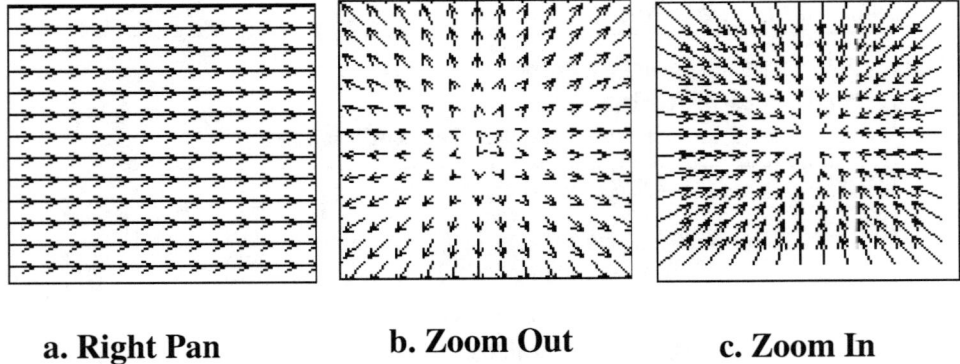

a. Right Pan **b. Zoom Out** **c. Zoom In**

Figure 3.7. Examples for motion flow types for camera motion.

still shot has a low intensity of activity while a "fast break" basketball shot has a high intensity of activity. This attribute can be expressed as a single integer lying in the range $[0, n]$, where n is 128, 256, or some other power of two integer. The semantics express the intensity of activity. The Activity Intensity was measured and then clustered on the following discrete scale: 1. Very Low Activity; 2. Low Activity; 3. Moderate Activity; 4. High Activity; 5. Very High Activity;

2. *Direction of Activity.* While a video shot may have several objects with differing activity, we can often identify a dominant direction. The Direction parameter expresses the dominant direction of the activity, if any. The direction of activity cannot always be determined since there might be several different objects moving in different directions. However, if there is in fact a dominant direction, then it is extremely useful to include. A dominant direction cab be expressed as an angle between 0 and 360 degrees, and an associated confidence measure.

3. *Spatial Distribution of Activity.* The spatial distribution of activity indicates whether the activity is spread across many regions or restricted to one large region. It is an indication of the number and size of "active" regions in a frame. For example, a talking head sequence would have one large active region, while an aerial shot of a busy street would have many small active regions. This attribute can be expressed as three integers and a floating point number to express the variance. The semantics expresses the size, number, and location of the active regions.

Section 3.4. Motional Features Representation and Analysis 67

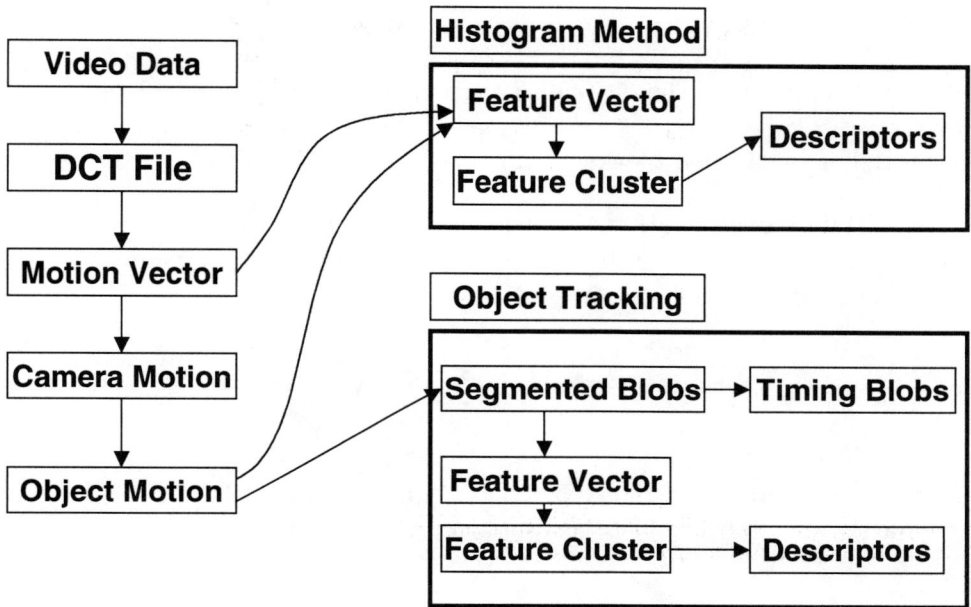

Figure 3.8. Object motion extraction flow chart for MPEG-1 video.

4. *Temporal Distribution of Activity.* The temporal distribution of activity expresses the variation of activity over the duration of the video segment/shot. In other words, whether the activity is sustained throughout the duration of the sequence, or whether it is confined to part of the duration. This attribute can be expressed as n integers, where n is 3, 4, 5, etc., as well as a floating point number to indicate the variance. The variance parameter functions as a confidence measure. However, accompanying every parameter with a confidence measure can be problematic for some applications. The semantics express the relative duration of different levels of activity in the sequence.

The Extraction Methods for the descriptor are as follows:
Intensity and Spatial Characteristics: The motion parameters based on motion vector magnitude for a frame. In Equations (3.4.1) and (3.4.4), $(x_{i,j}, y_{i,j})$ is the motion vector associated with the (i,j)th macroblock in dct images and we consider only P frames/objects. For each object or frame the

"activity matrix" C_{mv} is defined below.

$$C_{mv} = \sqrt{x_{i,j}^2 + y_{i,j}^2} \qquad (3.4.1)$$

and we can construct the descriptor for a frame in the following steps.

- 1. For intracoded blocks, $R(i,j) = 0$.

- 2. The average motion vector magnitude per macroblock of the frame or the object is given by Equation (3.4.2), where M, N are width and height in macroblocks, respectively.

$$C_{mv}^{avg} = \frac{1}{MN} \sum_{i=0}^{M} \sum_{j=0}^{N} C_{mv}(i,j) \qquad (3.4.2)$$

- 3. Compute the variance σ^2 of C_{mv} as Equation (3.4.3), where M, N are width and height in macroblocks, respectively.

$$\sigma_{fr}^2 = \frac{1}{MN} \sum_{i=0}^{M} \sum_{j=0}^{N} (C_{mv}(i,j) - C_{mv}^{avg})^2 \qquad (3.4.3)$$

- 4. Compute "run-length" features of C_{mv} as follows.

  (a) Use C_{mv}^{avg} as a threshold on C_{mv} as

$$C_{mv}^{thresh} = \begin{cases} C_{mv}(i,j), & C_{mv}(i,j) \geq C_{mv}^{avg}, \\ 0, & \text{otherwise.} \end{cases}$$

(b) Compute lengths of runs of zeroes in the above matrix using a raster-scan order. Classify the run-lengths into three categories: short, medium, and long that are normalized with respect to the object/frame width. In this case we have defined the short runs to be 1/3 of the frame width or lower, the medium runs to be greater than 1/3 the frame width and less than 2/3 of the frame width, and the long runs to be all runs that are greater than or equal to the width. N_{sr} is the number of short runs, with N_{mr} and N_{lr} similarly defined.

The parameters for a frame thus consist of C_{mv}^{avg}, σ_{fr}, N_{sr}, N_{mr}, and N_{lr}. The variance parameter expresses the intensity of activity. The run-length parameters express the spatial characteristics of the shot. Note that the run-length features indirectly express the number and size and shape of distinct moving objects in the frame and their distribution across the frame. For a frame with a single large object such as a talking head, the number of short run-lengths is high, whereas for a frame with several small objects, such as an aerial shot of a soccer game, the number of short run-lengths is lower.

Directional Characteristics: The motion vector direction-based parameters for a frame. For each object or frame the "angle matrix" A is defined as

$$Ang(i,j) = \tan^{-1}(y_{i,j}/x_{i,j}) \qquad (3.4.4)$$

The descriptor of directional characteristics for a frame is constructed as follows:

- 1. For intracoded blocks, $Ang(i,j) = 0$.

- 2. The average angle per macroblock of the frame/object Ang^{avg} is given by

$$Ang^{avg} = \frac{1}{MN}\sum_{i=0}^{M}\sum_{j=0}^{N}Ang(i,j) \qquad (3.4.5)$$

- 3. Compute the variance of $Ang(i,j)$ as Equation (3.4.6), where M, N are width and height in macroblocks, respectively.

$$\sigma_{fr\ r}^{ang2} = \frac{1}{MN}\sum_{i=0}^{M}\sum_{j=0}^{N}(Ang(i,j) - Ang^{avg})^2 \qquad (3.4.6)$$

The Motion Vector-Based Parameters for a Shot. The parameters for a shot are calculated using one of the following methods:

- a. Compute the descriptors of all the P frames in the shot. Get the average motion vector magnitude per macroblock on motion vectors for the entire shot. Use the frame with C_{avg} closest to this average.

- b. Find the average of all the parameters over the shot. Use the set of averages as the feature vector for the shots. In addition to the average of the shot parameters, we can also calculate the variance (or standard deviation) of the average and deviation parameters, as well as of the other parameters. The deviation parameters provide confidence measures for the average based descriptor parameters described earlier. Thus, we have four statistical parameters and three run-length features extracted using motion vector magnitudes and four statistical parameters extracted using the direction of the motion vectors.

- c. Find the C_{avg} for each P-frame in the shot. Find the median of the C_{avg} over all the P-frames and let the frame number of the median be n. Then use the descriptor of frame n as the shot descriptor.

- d. Choose a P-frame at random from the shot and use its descriptor as the shot descriptor.

Method b is our method of choice for the calculation of the shot descriptor parameters based on motion vectors.

Temporal Characteristics: Shot Activity Histogram. Let $A(t)$, $A(t+1), \ldots, A(t+n-3)$ denote the $n-2$ activity level values obtained for a shot with n frames. A shot activity histogram is defined as the normalized histogram of $A(t)$ values of that shot. Let A_{min} and A_{max} denote the minimum and maximum values of $A(t)$. Let H denote the shot activity histogram with b bins and $H(i)$ denoting the value at i^{th} bin. Then $H(i)$ is computed in two steps: 1. $H(i)$=number of $A(t)$ values between $i\frac{(A_{max}-A_{min})}{b}$ and $(i+1)\frac{(A_{max}-A_{min})}{b}$. 2. $H(i) = \frac{H(i)}{\sum_{j=0}^{b-1} H(j)}, 0 \leq i < b$.

3.5 Audio Features Representation and Analysis

Generally, there are two categories of audio features: time-domain and frequency domain. To extract out both domain features, we first sample audio signals at 11025 Hz and with 16 bits/sample. For each of the audio clips corresponding to a distinct visual scene, we then calculate audio features as follows.

3.5.1 Time-Domain Features

In the time domain, we calculate statistical parameters (such as mean, standard deviation, dynamic range, etc.) of trajectories of short-time audio volume and zero-crossing rate for each audio clip. Nonsilence ratios are also calculated based on both short-time volume and zero-crossing rate. The short-time audio volume and zero-crossing rate are defined in Equations (3.5.1) and (3.5.2), respectively.

$$V_n = \sqrt{\frac{1}{N} \sum_m [x(m)w(n-m)]^2}, \qquad (3.5.1)$$

$$Z_n = \frac{1}{2} \sum_m |sgn[x(m)] - sgn[x(m-1)]| w(n-m), \qquad (3.5.2)$$

where

$$w(n) = \begin{cases} 1, & 0 \leq n \leq N-1, \\ 0, & \text{otherwise}. \end{cases}$$

and

$$sgn[x(n)] = \begin{cases} 1, & x(n) \geq 0, \\ -1, & x(n) < 0. \end{cases}$$

In both the above equations, $x(m)$ is the discrete time audio signal with index of m, n is the time index of the short audio frame whose size is specified by a rectangle window of $w(m)$ with window length N. Here we choose frame size as $N = 150$ samples (i.e., the audio clip is about 15 ms long) and calculate both features once every 100 samples (about 10 ms apart) in the audio clips. The statistical parameters of the above two features, such as mean and variance, are calculated based on index n. Since the dynamic ranges of these statistical features differ a lot, we normalize them by their maximized volume and zero-crossing rate, respectively, for each audio clip. To figure out the silence ratio in each clip, we compare the volume and zero-crossing rate with a certain threshold for each. The audio frame claims to be silent when both its volume and zero-crossing rate are smaller than each of its thresholds. So the nonzero ratios of V_n and Z_n are the percentage of nonsilent frames over the whole audio clip.

3.5.2 Frequency-Domain Features

In the frequency domain, we first calculate the spectrum of an audio clip by using a direct FFT transform with a 512-point FFT size, which generates a 2D plot of the short-time Fourier transform (over each audio frame), with frequency as the x axis and amplitude in db as the y axis for all frames over the time domain. The following features are calculated for each frame and their distributions over the entire audio clip.

- *Short-time fundamental frequency.* Each frame's short-time fundamental frequency is calculated as Zhang and Kuo described in ([104]).

- *Short-time fundamental frequency distribution for the whole audio clip.* Statistical parameters, such as mean and variance, are calculated for the trajectory of the short-time fundamental frequency of each audio frame over the time through the entire audio clip.

- *Centroid frequency and bandwidth.* Similar to the proposal by Wold et al. ([97]), the frequency centroid, $C(i)$, and bandwidth $B(i)$, of each audio frame are defined as

$$C(i) = \frac{\int_0^\pi \omega \mid S_i(\omega) \mid^2}{\int_0^\pi \mid S_i(\omega) \mid^2}, \qquad (3.5.3)$$

$$C(i) = \frac{\int_0^\pi (\omega - C(i))^2 \mid S_i(\omega) \mid^2}{\int_0^\pi \mid S_i(\omega) \mid^2}, \qquad (3.5.4)$$

where $S_i(\omega)$ represents the short-time Fourier transform of the ith frame. Using this formula, we calculate centroid and bandwidth for every frame over the entire audio clip (remembering that each adjacent frame had overlap, as we chose frame size for 150 samples but calculated the frame every 100 samples), thus generating a 2D plot of the centroid and bandwidth along the time axis. The mean and standard deviation of both the centroid and bandwidth of an audio clip are used as four frequency domain features.

- *Energy ratio of some sensitive subbands.* The energy distribution in different frequency bands varies quite significantly among different audio signals. For example, the spectral peak tracks in speech normally lie in the lower frequency bands, ranging from 100–300 Hz; while whistles, which are often heard in sports videos, have high frequencies and strong spectrum energy with frequencies ranging from 3500–4500 Hz. To differentiate special audio events, like speech, whistles, or noise, we calculate the ratios of energy in the subbands [0–400 Hz], [400–1720 Hz], [1800–3500 Hz], [3500–4500 Hz] with respect to the overall energy for all frames in the audio clip.

- *Peak of spectrum on each frame and peak distribution of the entire audio clip.* The peak tracks in the spectrum of an audio signal often show us extra characteristic properties of the sound. In sports video games, one typical sound is the referee's whistle, which often occurs right after fouls in basketball and soccer or at the beginning of a serve in volleyball, etc.

Whistles in sports videos often last at least one second and have stronger energy than speech and music. So the whistle detection is linked to both video semantic boundary detection and semantic meaning inference. The peak of the whistle spectrum normally ranges from

3500 Hz to 4500 Hz. The most prominent frequency from FFT transformed spectrums is detected for every frame in an audio clip so as to get a 2D graph with the time domain on the X-axis. A whistle sound is detected if there is a longer than one second window of frequencies which fall into the range between 3500–4500 Hz.

3.6 Text Feature Representation and Analysis

The *vector space* ([80]) representation is widely used for information access in text documents. In this representation, each document is characterized by a Boolean or numerical vector. These vectors are embedded in a space in which each dimension corresponds to a distinct *term* in the corpus of documents being characterized. A given document vector has in each component a numerical value denoting some function f of how often the term corresponding to that dimension appears in the document. By varying function f, alternative term "weightings" ([79]) can be produced. Some standard weighting functions below are explored, and examples of vector representations of documents by using these schemes are also given.

The most common definition of a term (in English, as well as most other languages that use the Roman alphabet), and the one to use in all subsequent chapters, is that a term is a sequence of alpha-numeric characters which is delimited by white space (spaces, tabs or newline characters) or punctuation marks (such as a period or a comma). Moreover, all uppercase letters in a document are converted to lowercase so that capitalization is ignored. Consider the two short sample documents given in Tables 3.1 (quoted from Negroponte ([65])) and 3.2.

Table 3.1. Sample Text Document 1

> Computing is not about computers anymore. It is about living.

Table 3.3 shows the results of parsing these two documents into single-word terms, and then representing them as vectors with simple term frequencies (i.e., term counts) in each component. Such a representation is also referred to as a *bag-of-words* ([56]), since the relative position of terms in the

Table 3.2. Sample Text Document 2

> To live is to compute!

document, and hence the language structure, is not captured in the resulting vectors.

Table 3.3. A Simple Vector Representation of Two Sample Documents

Term	Vector for Document 1	Vector for Document 2
about	2	0
anymore	1	0
compute	0	1
computers	1	0
computing	1	0
is	2	1
it	1	0
live	0	1
living	1	0
not	1	0
to	0	2

3.6.1 Frequency-Based Vectors

As mentioned earlier, various functions may be applied to the frequency of term occurrences in documents to produce "weighted" document vectors with the vector space representation. Let $TF(t_i, d)$ denote the number of occurrences of term t_i in document d. Some function f to $TF(t_i, d)$ may be applied to produce the value for the ith component of the vector for document d. For example, for vectors in Table 3.3, we simply use the identity function $f(\alpha) = \alpha$ applied to the term counts TF. Another popular function applied to text document term frequencies (TF) is the $TFIDF$ weighting ([79], [78], [51]). In this scheme, not only are term frequencies (TF) in each document used as part of the weighting function, but they are also the inverse document frequency (IDF) of each term in the entire collection.

More formally, we have the following definitions.

- *Term Frequency*: Let T be the collection of all terms used in all documents. The term frequency of term t_i (ith vocabulary in T), denoted by $TF(t_i, d)$, is the number of times (frequency) t_i occurs in the document d.

- *Document Frequency*: $DF(t_i)$ is the number of documents in which term t_i occurs at least once.

- *Inverse Document Frequency*: $IDF(t_i) = log(\frac{|N|}{DF(t_i)})$, where $|N|$ is the total number of documents in the collection and $DF(t_i)$ is the number of documents containing term t_i.

- *Term Frequency × Inverse Document Frequency*: $TFIDF(t_i, d) = TF(t_i, d) \times IDF(t_i)$

$TFIDF$ weighting was widely used in the past primarily for text retrieval. It is also used to construct a classifier based on text cues for video classification in later chapters. a

3.6.2 Boolean Vectors

Alternatively, we may consider the use of a simple Boolean representation of documents by simply recording whether a given term appears in a document. In this case, we have

$$f(\alpha) = \begin{cases} 1 & \alpha \geq 1, \\ 0 & \text{otherwise.} \end{cases}$$

Table 3.4 shows the two sample documents in the previous section cast as Boolean vectors. Note that most rule-based methods ([3], [16]) are essentially based on an underlying Boolean model, as the antecedents of the classification rules they produce are only considering word presence and absence in documents. Thus, the Boolean vector representation of text information can also integrate into our rule-based video classification naturally.

3.6.3 Dimensionality Reduction Techniques

In applying the vector space representation to documents, it is clear that the resulting dimensionality of the space could be enormous, since the number

Table 3.4. A Boolean Vector Representation of Two Sample Documents

Term	Vector for Document 1	Vector for Document 2
about	1	0
anymore	1	0
compute	0	1
computers	1	0
computing	1	0
is	1	1
it	1	0
live	0	1
living	1	0
not	1	0
to	0	1

of dimensions is determined by the number of distinct terms in the corpus. Thus, methods to control the dimensionality of the vector space are needed. We show here how it is possible to use a few simple observations to significantly reduce the size of the feature space.

Word Stemming

In some cases, rather than defining terms to be distinct words in the corpus, word stemming is used to reduce words to some root form. Thus, terms that define the dimensions of the vector space are not actual words, but word stems. For example, words "computer," "computers," and "computing" would all be reduced to the word stem "compute." Porter ([73]) has developed a commonly used algorithm for word stemming, and this algorithm has been incorporated into the caption text parsing module of our system.

Eliminating Stop Words

There are many words in English that have little inherent topical content. These are words such as prepositions, conjunctions, and pronouns which are used to provide structure in the language rather than content. Such words are commonly used in documents regardless of topic, and thus have no topical relevance. We can therefore eliminate such words (and their dimensions corresponding to them) from our document vectors, as they will be of little use in clustering or classifying documents. Such words are commonly referred to as stop words, and their elimination from documents is common in text information retrieval ([76]). A list of exemplary stop words are: a, the, about, also, any, at, because, can, for, from, during, of, us, you, etc.

Multiword Terms

Some researchers have defined and reduced dimensions of a vector space by including multiword phrases. Examples of such multiword terms include phrases such as "President Bush" and "personal computer." Such multiword terms can be produced as a result of simply looking for frequently appearing sequences of words in the text documents ([17]). They can be used to detect certain meaningful and specific phrases for particular tasks.

Chapter 4

KNOWLEDGE-BASED VIDEO HIERARCHICAL CLASSIFICATION

4.1 Introduction

Efficient video classification can bridge the gap between low-level visual features and high-level semantic meanings, and thus facilitate video database indexing, visual data understanding, and machine interpretation of visual semantic meanings. Besides, video classification at different semantic levels can provide video summarization for fast browsing. However, video contains not only video signals, but also audio, speech, and text. These media display different characteristics, and express information at different levels and in different kinds of details. While trying to extract semantics from such data, it is necessary to utilize information available from all sources so as to improve the quality of the created metadata of video analysis and annotations. How to get the most useful semantic content from mixed media cues is an interesting and practical problem. In this book, a general video semantic abstraction concept called *the video semantic concept tree* with different levels of semantic granularities for video analysis and classification is described in detail. Here we use news video and general sports video as examples for concept illustration. In Chapter 5 we use them for experimental evaluation. Mixed media cues from visual, audio, and text information are used in a cooperative way so that video can be represented by both media low-level features and multiple high-level semantic concepts.

In this chapter we focus on the use of an inductive decision tree learning method to derive a set of if-then rules directly applicable to a set of visual and acoustical low-level feature-matching functions for video classification.

The rules for each video concept will be the trained knowledge. On the other hand, a $TFIDF$ text classifier, which will be described in Section 4.5.3, will be constructed independently to come up with a text vector rule for each concept of video data in the video semantic concept tree. There are several advantages to such a knowledge representation about video. First, these rules are powerful and easy to understand. Second, the computational complexity of classification based on these rules is low. Third, the rules are easy to integrate with a priori knowledge to provide more advanced reasoning. Finally, as video, audio, and text information are generally complementary to each other, combined rules from the video, audio, and text classifiers will improve final classification results.

The rest of this chapter is organized as follows. Previous work and common techniques for video classification and knowledge discovery are briefly reviewed in Section 4.2. The characteristics of video semantic contents such as video semantic layers are discussed and a video semantic concept tree scheme for general video is described in Section 4.3. The decision tree learning algorithm is described in Section 4.4. Knowledge-based video classifiers consisting of mixed media cues of visual, audio, and text information are described in Section 4.5. The on-line video classification system and the classification procedures are described in Section 4.6. Finally, a summary of the chapter is provided in Section 4.7.

4.2 Previous Work

Extensive research has been done over the years in developing cost-effective and performance-efficient low-level features for video indexing, while studies on video classification to get high-level semantic content have received increasing attention recently. For example, Jaimes and Chang ([43]) used an interactive learning algorithm to build a semantic model. In this system, they included users in the loop, allowed them to specify classes, provided them with examples for learning, and used learning algorithms to train classifiers. However, most video classification systems are feature-based for off-line applications and few of them have studied the relationship between low-level features and conceptual meanings. In an off-line video classification environment, classification results can be obtained based on all available low-level features, but this is usually not practical for real-time on-line video classification applications. Hence, a fast video classification system based on certain simple and easily extracted low-level features is essential. It would be even better if the system could guide feature extraction so as to avoid extracting unnecessary features, and thus saving the valuable computing power and reducing latency for on-line video content analysis.

Common classification techniques used in video/audio database applications include the Hidden Markov Model classifier, ([11], [67], [27]) the neural network classifier ([26], [46]), the decision tree learning algorithm ([107]) , the K-NN based classifiers ([90]) and the Bayesian network classifier ([61], [93]). These techniques exhibit these features. While the neural network classifier often hides internal mappings between classes and low-level feature attributes, it requires a whole feature vector for classification. The K-NN classifier does not need training, but it is much slower than the other classifiers if the vector dimension is high. The Hidden Markov Model classifier is good in classifying events with temporal progressive states, and has been widely used in speech recognition ([11]) and video/audio classification ([67], [27]). Most interestingly, the decision tree learning algorithm ([56]) is one of the most practical and widely used methods to generate rules in inductive inference. It splits classes by maximizing the "information gain," which constructs the final decision tree in a hierarchical representation naturally. Thus, the decision tree learning algorithm is used to construct the video/audio classification functions of the abstracted video semantic concept tree.

Video is a rich multistream source that contains audio, image, and text information, and each type of media stream expresses a certain level of semantic meaning. Text annotation has been used widely in visual database applications. For example, Johnson et al. ([47]) analyzed typical physicians' reports by using natural language processing techniques and a hierarchical terminology tree to annotate and index the medical images. Semantic information derived from visual data is, however, more direct and accurate. In this book, we present a method to extract and classify the semantic content in video by using mixed media cues for more extensive results. Some of the previous work is as follows. The Informedia project ([95], [94]) at the Carnegie Mellon University (CMU) was one of the six Digital Library projects sponsored by NSF. It created a terabyte digital video library (presently containing only news and documentary video) that allowed users to retrieve/browse a video segment by using both textual and visual means. A key feature of the system is that it combines speech recognition, natural language understanding, and image processing for multimedia content analysis. Text transcripts generated by speech recognition from the audio track are analyzed to select keywords for news video indexing and browsing. Maybury et al. ([53]) also used information from video and text to automatically segment video and analyze news video content for summarization and visualization. Huang et al. ([39], [40]) used mixed media cues of video, audio, and text to automatically segment news videos into different hierarchical levels, such as commercial breaks, news anchorperson, news summary, and news story sequences. However, the semantics, which Huang, et al. ([39]) extracted from videos, are still at a very high level of abstraction and thus often fail when people want to do more specific and sophisticated queries, such as "Please find me all basketball news related to the Lakers" or "Find me basketball game clips with dunks" or "Find me stories or close-ups of sports stars such as Michael Jordan." It is also more helpful for archiving and retrieving news stories if different granularity categories can be assigned to them.

Knowledge-based techniques ([19]) are widely used in the development of vision systems, such as image and video segmentation ([64]). Knowledge-based systems, which are also known as expert systems, have been traditionally used for high-level interpretation of images. They incorporate mechanisms for spatial and temporal reasoning that are characteristics of intermediate and high-level image understanding. However, to maintain their efficiency, knowledge-based systems are usually developed for specific applications.

In this work, we are interested in developing a knowledge-based system for fast on-line video classification and filtering applications. A generic framework will be established first. Then, we will classify and filter some CNN Headline News stories and NBC's Olympics 2000 sports game program, in particular, a basketball video, as examples to illustrate the concepts described in such a video semantic indexing system. Since video semantic structures and their perceptual patterns in visual/audio data are traceable, we did not include the user in the loop but instead used a knowledge-based classifier which is trained off-line. For example, in sports video, the rules for each sport are fixed because low-level features, such as motion and other visual features, will display certain fixed characteristics throughout the video as long as they belong to the same sport. This provides us with the possibility of establishing an off-line knowledge-based system for each sport in advance. Thus, in this book, we do not provide a classifier to classify any video's semantic meanings, but instead develop a generic method to classify video sequences into semantic units according to supervised learned rules. Since some semantically different video sequences might share the same low-level perceptual features, our goal is to find a generic way to determine those other features which can discern between any two videos that belong to semantically different categories. In addition, we will determine the priority, the order, and the weight of the features specified to differentiate between the two semantic video sequences.

4.3 Video Semantic Concept Tree

4.3.1 Characteristics of Video Semantic Content

The word "semantic" stands for "of or relating to meaning in a language." The semantics of video data can be thought of as the meaning of video and can be described by text descriptions. Since the amount of on-line data is increasing astronomically, the design of an efficient algorithm or approach to access the data has become of great interest and importance, especially for broadcast multimedia data. Besides, the need for querying with semantic key-words/key-phrases/key-concepts has motivated recent research in semantic video indexing. Creating and organizing a semantic description of the underlying multimedia data is an important step in achieving efficient discovery and access to relevant data. This is, however, severely hampered by the gap between the media low-level perceptual features and the high-level conceptual semantics. When we are trying to provide a solution to the problem of video semantic content analysis and organization, there are a few characteristics of video semantic content which are helpful for us to take into account.

First, semantic content is not easily defined for image/video because different people looking at the same visual data might pay attention to different content and arrive at different semantic values for the same video stream. For example, the general audience that watches basketball may care more about events such as scores, dunks, etc.; while a basketball team coach may pay more attention to a team's defense/offense strategies. Thus, to allow more generic information access, the provision of multivalue-specified semantic content analysis is very important. Second, because of the hierarchical and layered nature of video data, there is a need to classify video data at all possible levels, such as scene level, shot level, and object level. Third, semantic information itself has different levels of detail and granularity. For example, the concept of "vehicle" covers all the meanings of "car" and "truck," while "car" and "truck" include finer granularities of the concept "vehicle." Thus, to accommodate more application specifications, a hierarchical scheme con-

Section 4.3. Video Semantic Concept Tree   83

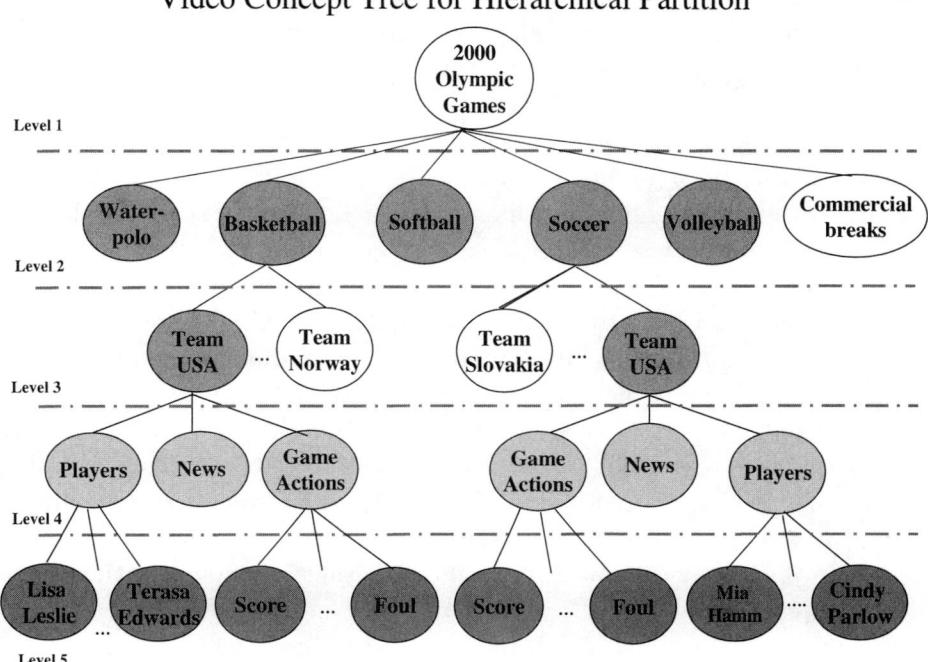

Figure 4.1. Illustration of the video concept tree.

sisting of different levels of video semantics and at different level of video structures is very necessary and useful. Fourth, object-based features, such as a segmented object from the background of an image and the features associated only with this object, are more likely to have a better semantic correlation. It is, however, not necessary that object-based features have better semantics. For example, in the case of motion pictures where there is no prominent object in the picture (such as a landscape video) or there are a lot of objects with similar motion patterns, a more global feature could be more useful to differentiate the two. Thus, different features are associated with different types of video semantic contents, and it is difficult to come up with one universal set of features which are satisfactory for expressing the semantics of all kinds of video.

4.3.2 Creating a Video Semantic Concept Tree

Based on our observations about the characteristics of video semantic content in the previous section, it is natural to construct a hierarchical video semantic concept tree to make an abstraction of the indexing keywords used in video applications, such as video database or video on-line filtering. Besides, a hierarchical semantic organization is able to accommodate more application specifications. For example, many Internet search engines, portals, and news websites, such as Yahoo![1] and CNN websites[2], also organize and store plain-text news hierarchically. Querying and filtering with respect to a concept hierarchy are more efficient and reliable than searching for specific keywords since semantic meanings of collected data are refined as we go down through the hierarchy. Also, querying can be more efficient when the hierarchy arranges concepts in some way that is meaningful to the user.

To link a semantic label with a representation of the observed visual data in a scalable way, we define a hierarchical concept tree which can be mapped to general video data based on each specific video type. The hierarchical concept tree is an extension of our published work ([107]), accomplished by generalizing the video data abstraction model and knowledge representation. A video hierarchical concept tree is supported by a set of labeled nodes called the concept hierarchy and a set of classification rules associated with multiple features at each node.

A concept hierarchy H is defined as a connected, acyclic, undirected, rooted tree, $H = (N, E)$, where N is a node (i.e., concept or class category in the hierarchy) and E is a binary relation between two nodes. A traversing from node N along the edge E upward will arrive at a parent node N', which represents a broader concept; whereas going downward will focus into more specific concepts, which are covered or included under parent nodes. Figure 4.1 shows a hierarchical concept tree that categorizes and abstracts sports video with semantics in five granularity levels: 1. sports broadcast programs; 2. sports types; 3. teams; 4. games, news, and players of each team; 5. special game events in each game or an individual player on the team.

Special game events and an individual player are often of special interest in many query problems because people like to know the performance and stories of sports superstars. Concepts shown in the concept hierarchy are meaningful to users and to their interests. Also, the video concept hierarchy is very general and suitable to all team sports videos since most sports games share similar scenarios. For example, every game is associated with two teams that are playing it, and all game events and news items are associated

---

[1] http://www.yahoo.com/
[2] http://www.cnn.com/

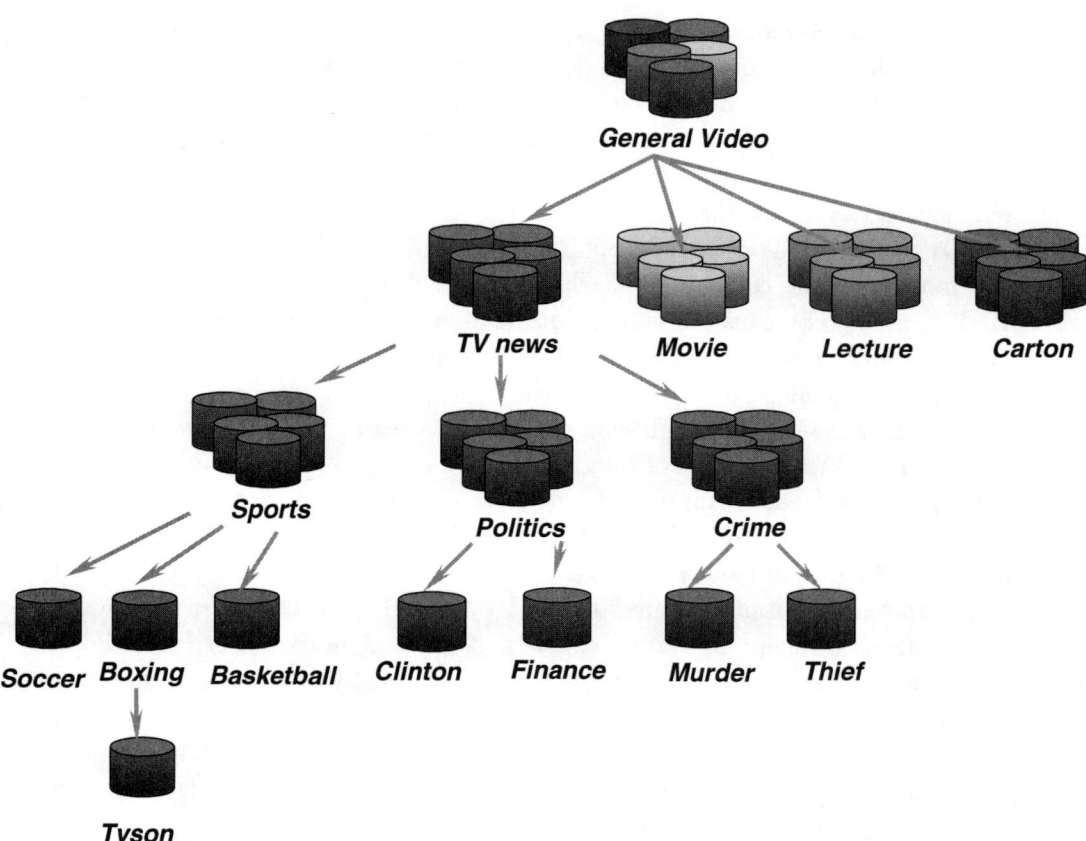

Figure 4.2. Illustration of a general concept tree.

with a team and its players. People often query about their favorite star players who often appear in close-up images and are often commented on by reporters. Since the concept hierarchy is modeled on semantics of a typical sports channel, and team sports videos follow a similar grammar, we can use almost 80 percent of the concept hierarchy for different team sports such as football, baseball, hockey, etc. While the overall scheme is applicable to all sports, it also contains a specific model (i.e., the domain knowledge) for each type of sport. For example, key events in basketball games are related to scoring and dunking – maneuvers which are different from those in soccer games. Thus, we manage to put domain knowledge in the concept tree, which will guide us in classifying video into its most suitable semantic categories. The concept tree can also guide users in video retrieval and filtering.

The semantic concept tree is a flexible and open-ended framework that explicitly accounts for the dependency, relationship, and co-occurrence between semantic concepts at various hierarchical levels. It is reasonable for us to relax the query constraints to the upper levels when the answer to a specific query on certain nodes is not available because every parent node exclusively covers its child nodes.

To construct a video concept tree for later classification and queries, we first define the concept hierarchy by using our domain knowledge. Note that it can also be constructed automatically by using unsupervised learning or clustering. Each node is accordingly labeled by a concept term. Video data are then collected according to the concept hierarchy. Another concept tree for hierarchical video classification is shown in Figure 4.2, which is an extension of Figure 4.1. Training and test data for video classification are used in supervised learning. An inductive decision tree learning algorithm is used to derive a model for future video data classification. According to the video hierarchy concept tree, a hierarchical classifier consists of a set of video and text classifiers that are used to classify a subtopic in a concept hierarchy. A concept hierarchy consists of a set of nodes where each node represents a specific topic. A topic is represented by a node in level n. It has subtopics in level $n+1$. Topics become more general (or more specific) as the hierarchy is traversed upward (or downward). Based on the concept hierarchy, we can determine a measure or a standard (i.e., the classifier) by which we can classify text and video data.

4.4 Decision Tree Learning Algorithm

Document classification involves assigning documents to one of several sets of *predefined* categories. In the classification task we are given a training set of data with preassigned class tables. From this data, we can then learn a model that can be used to classify new data into one of the existing classes.

Formalizing this notion, we begin with a set D_{train} of m training documents, denoted $d_1, ..., d_m$. Each such document also has associated with it a class label, denoted by the variable C. Learning a classifier, also known as *supervised* machine learning (since we begin with a set of prelabeled data), involves inducing a model from the training data that we believe will be effective at predicting the class label C of new data for which we do not know the class. This new data is often referred to as the testing set D_{test} in experimental evaluations.

Having previously discussed a general scheme showing how to assign predefined class labels for common sports videos, we now turn to the question of how to induce a classification model from a prelabeled set of data. First, we provide a brief introduction to the decision tree induction algorithm which we will use to learn the video/audio classifier model for our system.

Decision tree learning is one of the most widely used and practical methods for inductive inference. It is a method for approximating discrete-valued target functions, in which the learned function is represented by a decision tree. The algorithm searches a completely expressive hypothesis space, thus avoiding the difficulty of a restricted hypothesis space. It can also be represented as a set of if-then rules to improve human readability.

Several algorithms have been studied for inducing decision trees, including the CART algorithm of Breiman et al. ([8]), Quinlan's ID3 algorithm ([74]), and its more recent incarnation as C4.5 ([75]). Bayesian methods for constructing decision trees have also been proposed and studied by Buntine ([9]). We focus on the C4.5 algorithm since it is the most widely used tree induction method in the machine learning community.

C4.5 induces a decision tree in a greedy divide and conquer fashion. As the tree is being constructed, each instance attribute is evaluated by using a statistical test to determine how well it alone classifies training examples. The best attribute is selected and used as a test at the root node of the tree. A descendant of the root node is then created for each possible value of this attribute, and training examples are sorted to the appropriate descendant node (i.e., down the branch corresponding to the example's value for this attribute). The entire process is repeated by using training examples associated with each descendant node to select the best attribute to test at that point in the tree. This forms a greedy search for an acceptable decision tree in which the algorithm never backtracks to reconsider earlier choices. The choice of which feature to split on at a given node is made by selecting the feature which maximizes a certain criterion. To explain this, we first define a few parameters as follows.

A statistical property, called the *information gain*, is defined below. Given a collection of S, containing category 1 and category 2 of some target concepts, the entropy of S relative to this Boolean (category 1 or category 2) classification is:

$$\text{Entropy}(S) = -p_1 \log_2 p_1 - p_2 \log_2 p_2. \qquad (4.4.1)$$

More generally, if the target concept category attribute can take on c different values, the entropy of S relative to this c-wise classification is defined as

$$\text{Entropy}(S) = \sum_{i=1}^{c}(-p(i)\log_2 p(i)), \qquad (4.4.2)$$

where $p(i)$ is the proportion of S belonging to class i. The entropy is equal to 0, which is the minimum, when all cases in a set belong to the same class while the entropy is equal to 1, which is the maximum, when each class is equally distributed in the given set.

The *information gain* is simply the expected reduction in the entropy caused by partitioning examples according to certain feature attributes. More precisely, the information gain $Gain(S, A)$ of an attribute A relative to a collection of examples S is defined as

$$\text{Gain}(S, A_v) = \text{Entropy}(S) - \sum_{v \in Values(A)} (\frac{|S_v|}{|S|} Entropy(S_v)). \qquad (4.4.3)$$

$Gain(S, A)$ can also be expressed as

$$\text{Gain}(S, A_v) = -\sum_{i=1}^{c}(p(i)\log_2 p(i)) + \sum_{v \in Values(A)}(p(i, A_v)\log_2 p(i|A_v)), \qquad (4.4.4)$$

where $Values(A)$ is the set of all possible values for attribute A, S_v is the subset of S for which attribute A has value v, and $Gain(S, A)$ is the information gain provided about the target function value, given the value of the attribute of A. The value of $Gain(S, A)$ is the number of bits saved when encoding the target value of an arbitrary subset of S by knowing the value of attribute A. In other words, the $Gain(S, A)$ is a measure of the reduction in entropy of the class variable C after the value for the feature A is observed.

Besides the $Gain(S, A)$, a $Gain\ Ratio(A)$ is defined as

$$\text{Gain Ratio}(S, A) = \frac{Gain(S, A)}{splitinfo(A)}, \qquad (4.4.5)$$

where the denominator in Equation 4.4.5 is given by

$$\text{splitinfo}(A) = -\sum_{v \in Values(A)} p(A_v) log(A_v), \qquad (4.4.6)$$

which simply represents the entropy in the variable A, which can alternatively by seen as the reduction in entropy realized by selecting this variable to split on.

From Eqs. 4.4.4 and 4.4.6, it becomes clear that *Gain Ratio* is simply a measure of the proportion of information relevant to the class variable C derived from splitting on a given feature A. The whole tree induction process depends upon the split criterion and the stop criterion. A good tree should have fewer levels as it is better to classify with fewer decisions. Also, a good tree should have a large leaf population since the number of leaves represents the number of classification cases. In a learning algorithm, we split training data upon the variable that maximizes $Gain(S, A)$ or $Gain\ Ratio(S, A)$, where the "gain" value only depends on the class distribution, which makes the computation easily performed. Quinlan explains that the *splitinfo* function was introduced in Eq. 4.4.5 as a means of preventing the *Gain* function from overly favoring attributes with many values. He also reports that using *Gain Ratio* as a splitting criterion often empirically outperforms the use of *Gain* alone, even on entirely binary data.

4.5 Rule-Based Knowledge Base Construction

In order to classify incoming video streams into meaningful semantic classes, we should classify video with a small number of features that can be easily extracted. Figure 4.3 illustrates the overall system for knowledge-based building and on-line video classification. It consists of two steps. First, it performs the off-line training. Sample video clips of different categories are identified, and appropriate low-level features are created. Then, an entropy-based inductive tree-learning algorithm ([75]) is used to establish the trained knowledge base. Rules are learned by training, which includes ways to determine characteristic features for each class, the function of feature values, and the feature order and weight for each class. The algorithm can also update the knowledge, if necessary. Once we have the knowledge of rules for each class, they are used to build on-line feature extraction in response to filtering specified by users, i.e., choosing the appropriate operators to get target feature descriptors, and output the binary classification result (yes/no) to the user's specification. The rules can specify the order, the threshold value, and the priority of each feature test. Classification is hierarchical and automatic, which meets the fast on-line classification requirement. The above process will be described in detail in the next few sections.

4.5.1 Video Classification Rules in the Knowledge Base

Video concept classification functions can be stored in the knowledge base. It contains two parts: the first is the concept hierarchy specified for the in-

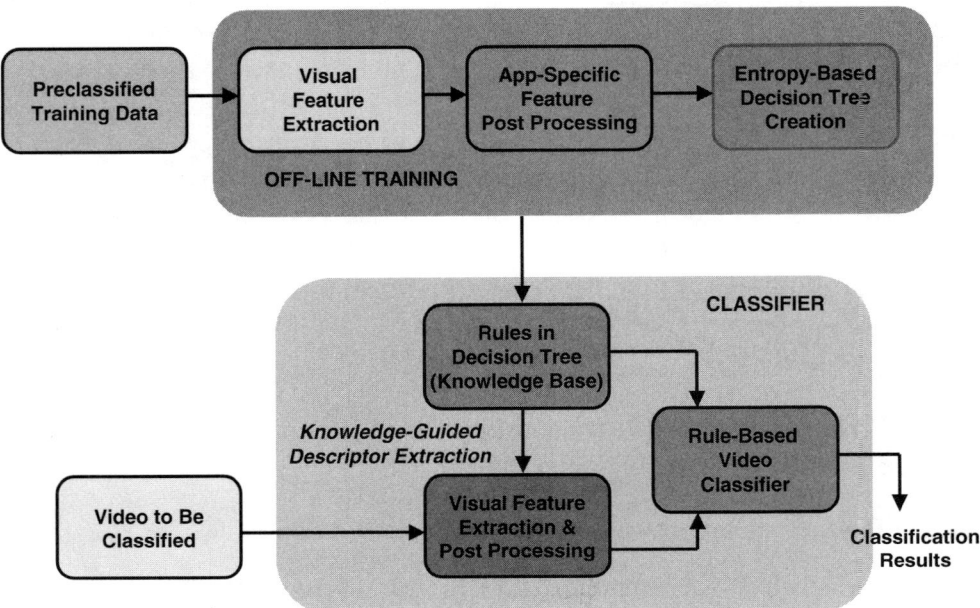

Figure 4.3. Knowledge base building and on-line video classification flow chart.

tended video conceptual contents; the second is the set of video classification functions for each concept.

Our knowledge base for video classification functions is represented as a decision tree, as shown in Figure 4.4, where each node in the tree is an if-then rule applied to a similarity metric based on appropriate low-level features along with well-derived thresholds. The rule is depicted as $f = < F, \theta >$, where F denotes the appropriate feature and θ the threshold which is automatically created during the training process. Semantic categories form the leaves of the decision tree. Each node in the tree is either a leaf or an intermediate node with two children. A set of videos are associated with each node N_i while a decision rule f_i is associated with each intermediate node.

An example of a rule-based tree with nine nodes is shown in Figure 4.4. The entire set of video concetps is associated with the root. Let N_i be an intermediate node, with its children labeled as N_{i1} and N_{i2}. The video subsets V_i, V_{i1}, V_{i2} satisfy:

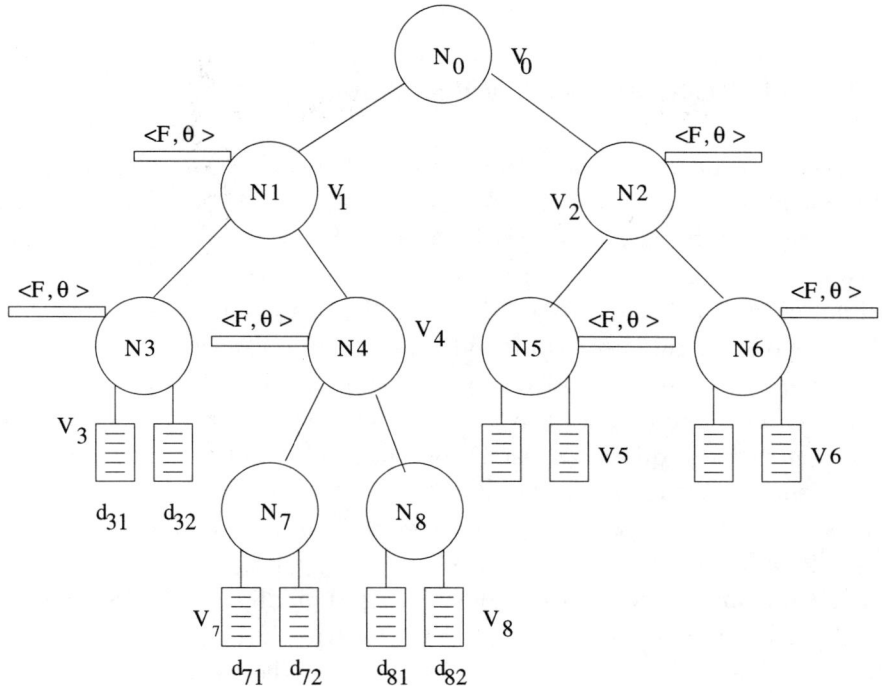

Figure 4.4. Illustration of a rule-based tree.

$$V_i = V_{i1} \cup V_{i2}, \; V_{i1} \cap V_{i2} = \emptyset$$

In other words, leaves form a partition of the database into disjoint subsets. In the example above, sets V_3, V_5, V_6, V_7, and V_8 are disjoint, forming the leaves of the whole tree, and their union is V_0. Thus, N_3, N_5, N_6, N_7, and N_8 represent the five concept categories in the tree.

Without loss of generality, it is assumed that the decision rule f_i is a discriminate function with the following interpretation. Let x denote a video clip in V_i. If $f_i(x, F_i) \leq \theta_i$, then $x \in V_{i1}$. If $f_i(x, F_i) > \theta_i$, then $x \in V_{i2}$. When these discriminating functions represent important visual characteristics, a visual-content rule tree partitions the set of videos into distinct clusters with feature-similar video clips in each cluster. In the example above, all images in V_7 have the following properties in common: $f_0(x, F_0) \leq \theta_0$, $f_1(x, F_1) > \theta_1$, $f_4(x, F_4) \leq \theta_4$. They should look different from the video set in V_2 if the characterization according to the rule of $f_0(x, F_0)$ is visually meaningful. In other words, three rules at most are needed to classify a video into class N_7.

4.5.2 Video Classifier with Video/Audio Cues

The classification scheme uses the rule-based system to build a binary tree and associates tree nodes with subclasses of videos. This enables fast video classification. The key computational step is to create the two children of a node so that their associated classes are clustered with respect to a meaningful visual characteristic. This implies that video subsets associated with each one of the two child nodes are more alike with respect to the visual property than the video clips associated with the parent node. If the tree is deep enough, its leaves should correspond to clusters of similar video events, and a query by visual-content or classification based on visual features should return one or more of these clusters of video.

First, the entropy-based decision tree learning algorithm ([75]) is used to get classification functions for each node by using video/audio features ([107]). Contrary to text feature node aggregation, all classification functions are collected from top to bottom. Classification functions of parents are passed to all children under them. The decision tree learning procedure for the whole concept tree is as follows:

1. The nodes from the same level of concept tree S are treated as a set of unclassified data. Nodes are split from the same level into subgroups upon the variable feature $<f>$ that maximizes the entropy gain ([75]). If S contains one or more tuples labeled by class C_i and then the decision tree is labeled as a leaf identifying class C_i, go to Step 4.

2. Otherwise, S contains tuples with mixed classification. S is split into $S_1, S_2, ... , S_m$ which all "tend to belong" to the same class. The split is executed according to possible outcomes $\{O_1, O_2, ... , O_m\}$ of a certain feature attribute r_k. Thus, S_i contains all $r \in S$ such that $r_k = O_i$. In this case, the tree for S is a node with m children. The node is labeled with feature attribute r_k and function $f = <r_k, O_i>$.

3. Perform the above two steps recursively for each $S_1, S_2, ..., S_m$.

4. Check if the nodes are in the bottom of the concept tree or not (in our experiment with the concept hierarchy in Figure 4.2, check if the node is at level five concept. If yes, stop. Otherwise, for each node, repeat the following. Go to the next level of concept nodes and pass classification functions to all its children and then go to Step 1.

Note that this procedure can also be extended to text feature representations. Since text feature vectors are normally very big for a large collection of text documents, the procedure could generate a very long list of rules for classification. However, the lowest-level nodes on a video concept hierarchy, such as the Level 5 concepts in our Olympic Games tree, are highly domain-focused, and thus the text vector feature can be shortened by incorporating background knowledge and selecting features deliberately to get rid of unimportant terms in the vector space of the nodes on that semantic concept level. Once the text feature vector dimension is reduced, we can train the decision tree learning algorithm with all the mixed media cues in video, audio, and text to get classification rules directly. This is especially useful for our events detection and classification, as some events are always accompanied by the same keywords by the commentators. One example is the word "goal" in soccer games. Whenever there is a successful score, a commentator will always emphasize it by announcing "Goal!"

4.5.3 Video Classifier with Text Cues

Today, many televised programs have closed captions, making them convenient for hearing-impaired people. In the meantime, captioned text provides us with an extra level of semantic interpretation of the visual data. Many researchers have used keywords from the captioned text to guide video segmentation and summarization ([53], [63]). Indeed, the area of free text understanding, classification, and summarization is itself a very active research topic. However, since a natural language has so much variation in expression, the use of a few key words for video indexing and retrieval limits the scope of applications and thus is not flexible and efficient enough. Nevertheless, for us, establishing a hierarchical concept from the captioned text point of view is meaningful, as we have established a counterpart from the visual/audio side. To convert a collection of captioned text documents in each node into effective and efficient representation, it is better to use a "bag-of-words" to represent a category of text documents and translate keyword-queries onto key word-vector matching in this sense. First, each document is preprocessed by using the *stopping* and *stemming* algorithms ([78], [51]), where stopping is the procedure to eliminate common words from the text, and stemming is the procedure to find a unique representation (e.g., root) for a word. After these procedures, the following text features for each text document corresponding to a video sequence are calculated:

- *Term Frequency*: Let T be the collection of all terms used in all documents. The *term frequency* of the term t_i (ith vocabulary in T) denoted by $TF(t_i, d)$ is the number of times (frequency) t_i occurs in the document d.

- *Document Frequency*: $DF(t_i)$ is the number of documents in which term t_i occurs at least once.

- *Inverse Document Frequency*: $IDF(t_i) = log(\frac{|N|}{DF(t_i)})$, where $|N|$ is the total number of documents in the collection and $DF(t_i)$ is the number of documents containing term t_i.

- *Term Frequency × Inverse Document Frequency*: $TFIDF(t_i, d) = TF(t_i, d) \times IDF(t_i)$.

All child text feature vectors are subsequently merged to obtain the TF vector at a node. The computation of TF vectors continues from bottom to top until the root is reached. In the meantime, the DF vector for each node is also propagated all the way up to the root. When this merging process is completed, the feature vector in each node is given by

$$\vec{F} = <TFIDF(t_1, d), TFIDF(t_2, d), \ldots, TFIDF(t_{|T|}, d)>,$$

where d is the union of documents (all children below the node are included) belonging to the node, and t is the union of the set of words in d.

The video classifier with text cues is represented with all the text feature vectors in the concept tree. However, to find similarities of new documents (n) to the text vector representation of each concept node (c), we calculate the projection of new document's text feature vector ($\vec{n}$) over that of the concept node($\vec{c}$), which is a *cosine* function. The closer the two vectors, the bigger the value of the projection. To admit a new document into a concept tree, we have to let the projection value be larger than a threshold θ which is trained to be the best value to include all the trained documents under that concept node. This statement is mathematically expressed as follows:

$$\cos(\vec{n}, \vec{c}) = \frac{\vec{n} \cdot \vec{c}}{\|\vec{n}\| \cdot \|\vec{c}\|} \geq \theta \qquad (4.5.1)$$

The training procedure to get a hierarchical classifier can be viewed in two phases. First, for each prototype vector $\vec{c}$, we need to introduce a *threshold* θ – a distance measure to indicate at what distance range to the prototype vector we consider documents fall into the same category (i.e., class). This becomes clear if we imagine that a $TFIDF$ vector is lying in the n dimensional hyper-space shown in Figure 4.5. As can be seen, the threshold makes a boundary for a class. For simplicity, we visualize it after normalization (i.e., as a unit-length vector) in a 3-D coordinate system. Second, we need to examine how good the hierarchical classifier is. That is, given the prototype vectors and thresholds, what is the *accuracy* with which the classifier will correctly classify documents? Later, we will check the accuracy for other brand-new test documents.

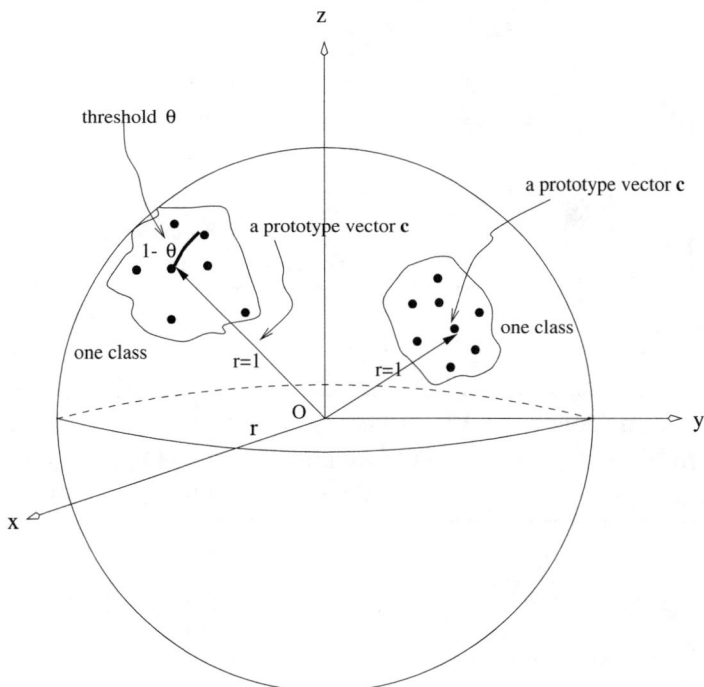

Figure 4.5. Learning the threshold to admit new documents into a class.

Training data are used to recursively adjust θ for every node so that each training document can get the largest similarity match with the text vector of the concept node. After training, a classifier is obtained for each node with a text classification function $cos(\vec{n}, \vec{c}) = \frac{\vec{n} \cdot \vec{c}}{||\vec{n}|| \cdot ||\vec{c}||} \geq \theta$, together with the video classifier with visual/audio differentiate functions described in the previous subsection.

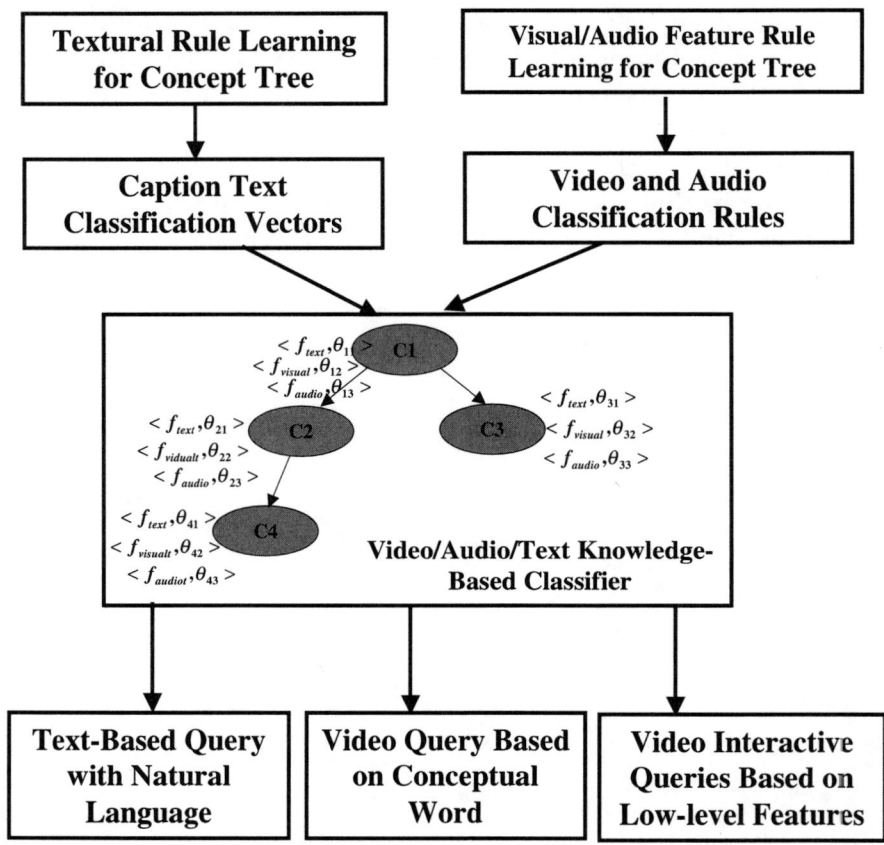

Figure 4.6. The classifier with mixed media cues.

After learning a video classifier for each node with visual/audio cues analyzed, independently from text cues, constructed classifiers with all the classification functions are labeled for the whole concept hierarchy tree as demonstrated in Figure 4.6. Classification functions $< f_i, \theta >$ are represented as a function of media features f_i from text/video/audio and the feature's threshold θ at each node. An example of possible classification functions with video/audio cues of video data for node C_3 in Figure 4.6 would be: if $f_{visual}(x, F_{32}) \geq \theta_{32}$ and $f_{audio}(x, F_{33}) \geq \theta_{33}$, then $x \in C_3$. Here, F_{32} and F_{33} are visual/audio descriptors and x is any of a set of unclassified audio/video clips.

4.6 On-Line Knowledge-Based Video Classification System

The on-line video classification system for filtering based on an on-line user's profile or an off-line query-based retrieval is shown in Figure 4.7. It is composed of four major modules: Media Browser (MB), Media Matcher (MM), Media Repository (MR), and Media Planner (MP). The client uses MB to specify their interests using a text or concept hierarchy, and/or to identify relevant data sources to be requested. The MM is the multimedia classifier that matches real-time audio and video streams broadcast over the Internet with predefined concept hierarchies, and stores related audio and video streams' pointers within proper nodes in the Media Concept Hierarchy (MCH) tree. Here, MCH refers to a video semantic concept tree. MP matches users' requests with MCH and builds a personalized execution plan to retrieve the relevant data sources for each user's profile/interest. The plan includes routines to analyze video/audio/text features to classify incoming streams; it also includes a description of how to stream and synchronize video and audio from multiple remote sources. MR reorganizes the knowledge base to enhance the semantic description (meta-data) based on new requests and new data streams. At the core of MR is the knowledge-based database of the system, which was described in detail in the previous section. In this section, we discuss MP and MM modules to describe on-line video classification procedures for filtering and querying tasks.

4.6.1 Media Planner (MP)

The Media Planner (MP) parses and matches users' requests from their specific queries/profiles with Media Concept Hierarchy (MCH), which refers to video the semantic concept hierarchy tree, and extracts relevant metadata describing the data sources that have contents matched to the user's requests. MP takes users' requests/profiles as the input, produces a script file containing commands to locate/retrieve data sources, and presents the data to the user. The media request in query takes one of two forms: a list of keywords (and/or short paragraphs) provided by the user, or a subset of paths from the Media Concept Hierarchy (MCH). The Semantic Request Parser (SRP) converts the specified subset of paths into a set of keywords and performs the matching with MCH.

The MM module described below trains a predefined MCH by using vector-based score of close-captioned data or decision-tree-based rules. Consequently, the hierarchy in MCH possesses knowledge about which class a new video data stream should belong to, and it helps in matching users' profiles/query inputs as a set of keywords with MCH to locate relevant data sources.

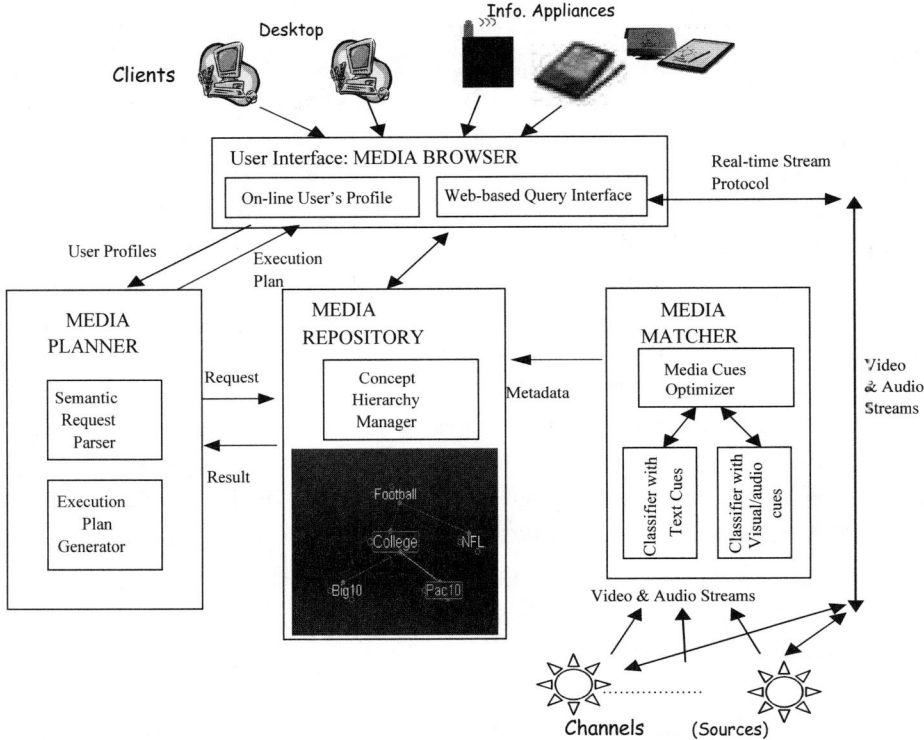

Figure 4.7. Illustration of a knowledge-based video classification and query system.

4.6.1.1 Classification Execution Plan Generation for Visual/Audio Cues

Learned decision trees are constructed with a top-down approach, beginning with the question: Which attribute should be tested at the root of the tree? The central functions of the video classification algorithm are to determine which attribute is the most useful in classifying examples and what is a good quantitative measure of the worth of an attribute. The goal of the algorithm is to find a value which can measure how well a given attribute separates training examples according to their target classification. The properties described above are very important in video classification since there are so many possible features to be used as keys to query video/image databases, and each feature is not always the best for all queries under different semantic environments. To solve the feature indexing and classification problem efficiently, the study of feature effectiveness for a certain classification application is essential.

The rule tree provides the optimal procedure to find a value that can measure how well a given attribute separates training examples according to their target classification. A new video clip is then classified as follows. Following the tree, the feature to be utilized in Level 1 (the root level) is first extracted and the corresponding rule is applied. The following path is then selected. At the next level, the same step is carried out whereby an appropriate feature is selected, and the corresponding rule applied. In this system, only those features that are relevant are extracted and they are matched with the rule threshold directly. Further processing, such as data indexing/filtering, will be made right after the classification match is done by MM.

The classification execution plan based on the visual/audio cues in a knowledge-based video classification/inference system is generated as shown in Figure 4.8. It consists of the following three major steps:

- *Preprocessing for Feature and Feature Extraction Selection.* Based on the target video concept specified in a user's profile or query, the system will search for nodes in the rule tree in the knowledge base Media Concept Hierarchy (MCH) by traversing up and down the nodes in the rule tree. If a concept keyword is identified in the video semantic hierarchy tree by MCH, relevant visual/audio features and their thresholds will be selected. At the same time, the feature extraction operators for the corresponding feature descriptor will be decided. If a user specifies queries or profiles using short paragraphs, classification execution plan generation will be based on text features which will be described shortly.

- *Knowledge-Based Content Matching.* Once each feature extraction operator obtains the feature value, it is compared with the rule by MM. If it matches, then the next feature extraction and matching operators are selected. The procedure is continued until the leaf node in the binary tree is reached.

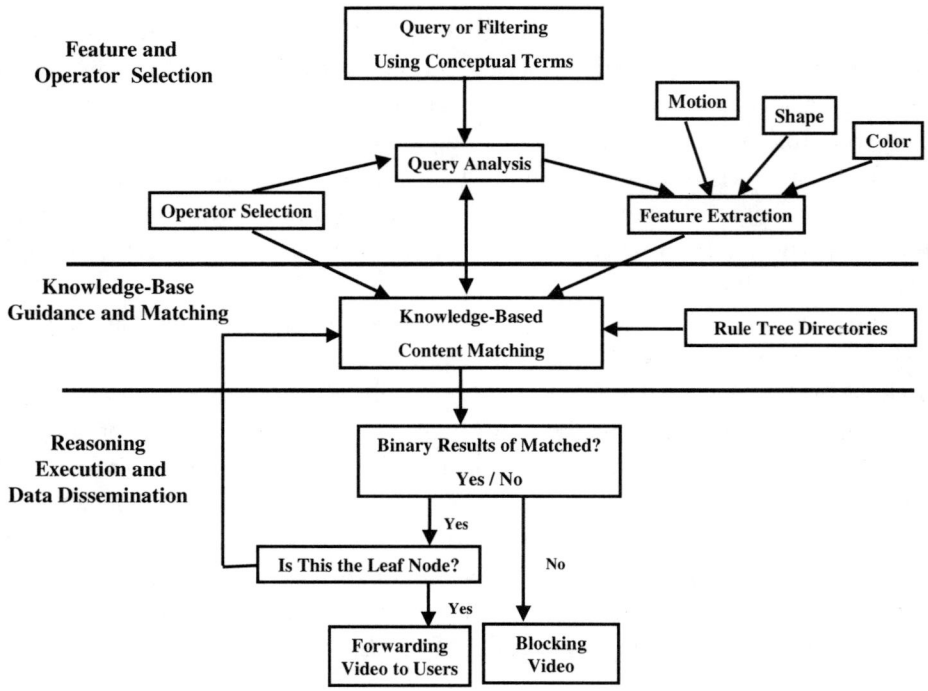

Figure 4.8. The flow chart of knowledge-based video classification.

- *Video Dissemination Based on Video Classification.* The raw video data, including audio and texts, are then sent out to (or blocked from) users if they are matched (or mismatched) with a user's profile or query requests.

4.6.1.2 Execution Plan Generation (EPG) with Text Classifier

Considering the input of user's request in the profile or query as a set of keywords or natural sentences/paragraphs, The Semantic Request Parser (SRP) will treat it as a close-captioned input to MM. Therefore, SRP converts the request input to a $TFIDF$ vector and searches for the target class in the video semantic hierarchy tree in Media Concept Hierarchy (MCH). Since any class (node) in MCH close enough to the input will be selected, the target class may not be limited to one. Once a video concept tree node is identified, a classification execution plan based on visual/audio cues can be generated based on our previous discussion, and a different classification execution plan with text cues only can be generated by MP with the pseudocode given in Figure 4.9.

Let **t** = TF(Keyword Input)
Class = ∅
NodeSet = root
while (NodeSet ≠ ∅) **do**
begin

    Retrieve a node N_i from NodeSet: NodeSet = NodeSet - N_i
    Retrieve the prototype vector: $\vec{c} = TFIDF(N_i)$
    Retrieve the threshold: $\theta = \text{Threshold}(N_i)$
    Compute the $TFIDF$ of user input: $\vec{n} = IDF(N_i) \times \mathbf{t}$.
    Compute the similarity between $\vec{n}$ and $\vec{c}$: s = $cos(\vec{n}, \vec{c})$.
    if (s ≥ θ)
        Class = Class ∩ {**c**}.
        NodeSet = NodeSet ∩ {GetChildren(N_i)}.
end
return Class

Figure 4.9. Pseudocodes used in search of the class concept node of MCH and generation of classification execution plan with text cues.

In the procedure, the user input does not have an IDF value, but borrows it from the node in the concept hierarchy. IDF multiplied with TF becomes a $TFIDF$ vector, making it comparable to the prototype vector of the video semantic concept hierarchy tree in MCH. After successfully matching users' profile/query with contents described by MCH, the Execution Plan Generator (EPG) generates the script of how to locate and retrieve contents from different sources from nodes in MCH.

4.6.2 Media Matcher (MM)

MM listens to live broadcasts from multiple sources over the Internet or TV channels. It uses a classification function of text cues to classify closed-captioned contents (audio and text) into relevant nodes in the hierarchy. Moreover, it will also use a set of classifications of visual/audio cues to classify video contents into the concept hierarchy. Note that text features are quite independent of audio/visual features, and classification functions based on different cues are constructed independently. Theoretically, a video can be classified based on classification functions from a single cue, such as text or visual/audio features alone. However, the details of text, audio, and visual contents, which are contained simultaneously in video streams, are complementary to each other; therefore, it is better to use all the cues in video to get optimized results.

There are two classification modes of operation. In the first one, the video classifier with the text cue and the video classifier with visual/audio cues can work independently to classify incoming streams into relevant nodes in the concept hierarchy. The second mode of operation will improve the performance and the accuracy of classification by providing a Media Cues Optimizer. Indeed, the Optimizer uses cues from texts during the text classification phase to trigger the video classifier at an appropriate time instance, or to make additional assumptions about video, and vice versa. A set of heuristic rules is used to describe the mixed media cues, and is domain dependent. For example, by using the text classification of a football game, the system knows that there is a touchdown at time t, and assumes that the corresponding video stream for the closed-caption will be approximately the n-th frames from time t. Therefore, the system can identify the exact video scene which shows the touchdown without performing additional video processing tasks. The output of MM will be MCH, and each node in the hierarchy will contain information such as URLs, Start Time, and Stop Time. MM stores the hierarchy back in MR.

In Figure 4.2, we also notice that classifiers with the text cues or the visual/audio cues perform differently at different levels of nodes in the hierarchical concept tree. Certain classification functions with one cue are obviously superior to their counterpart with different cues. For example, to classify team players, text cues are superior to visual/audio cues in terms of classification functions. For concepts in this case, heuristic rules are used in classification EPG to choose better classification functions automatically. By allowing queries based on either the text classifier or the visual/audio classifier, any inadequacy or limitation of a classifier from a single type of signal can be compensated. More complicated video classification and synthesis techniques to fuse the text classifier and the video/audio classifier could be constructed by using the Bayesian Networks and its performance needs further study.

4.7 Summary

We have described an integrated knowledge-based video classification system with mixed media cues for sports video in this chapter. By using a predefined video concept tree, the domain knowledge by supervised learning is enforced, so that the video classification system can pick up classification functions automatically for future intelligent video analysis. The system prototype is suitable for both on-line and off-line information access applications. One particular contribution of our work is automatic fusion of information from multiple media streams into a hierarchical video semantic structure, which gives a representation of video from the two sides of media low-level features and high-level semantic meanings. By incorporating mixed media cues and effectively coupling low-level features with high-level knowledge in the form of rules, our goal of video semantic understanding is realized. Experimental results will be discussed in Chapter 5.

Chapter 5

EXPERIMENTAL RESULTS

In this chapter, experimental results based on the system described in previous chapters are presented. In Section 5.1, we give results of CNN news decomposition and its table of contents generation. CNN news video is decomposed into useful story units for on-line filtering. In Section 5.2, segmentation of NBC 2000 Olympic game video is described. Experimental results on basketball event classification and hierarchical video classification based on the concept tree are presented. In Section 5.3, a comparison is made between our system and other related work.

5.1 CNN News Video Segmentation and Indexing

Because video stream distribution on semantic multicast requires semantic filtering instead of low-level visual feature filtering, the goal of our news video analysis and segmentation is to decompose the CNN news video into story units and to understand the semantic content of each story unit for news indexing and filtering. On-line video segmentation and story boundary detection require fast and efficient classification algorithm. To avoid generating large latency caused by the classification proxy, we consider a trade-off between the complexity of the classification algorithm and the speed (e.g., through binary classification). Since it is extremely difficult to classify video of unknown type and content, we would first learn the interest of users from user's requests or by supervised learning from the training data set, and then establish the knowledge base containing information about categories and the "powerful low-level visual features" that differentiate interesting classes based on the application domain knowledge.

5.1.1 TV News Video Production Rules

A TV news program is usually organized as a sequence of independent subjects which are introduced by a speaker called the anchorperson (Figure 5.1).

CNN Headline News Structure

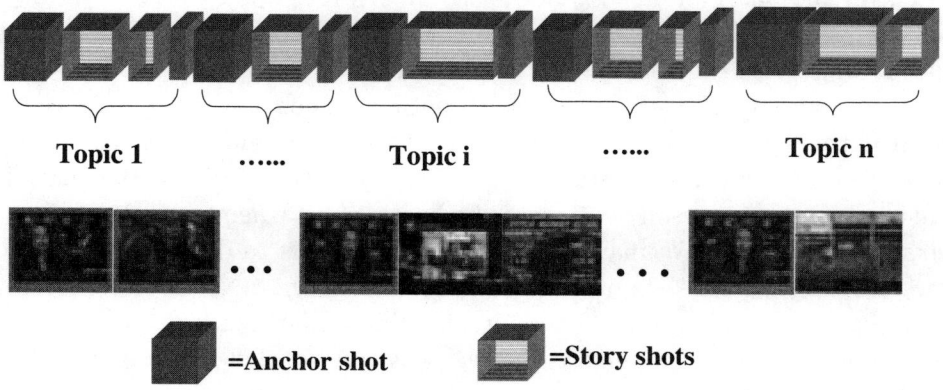

Figure 5.1. CNN news video structure for decomposition.

Each subject has an average duration between one and three minutes, and is composed of several shots. A text stream, which corresponds to the text transcription of the sound channel, can also be mapped in real time with the news program, as in many U.S. TV networks. Opposite to fiction movies, where there are many syntactic rules, most news programs follow very precise rules. For example, the station icon image will almost always appear before the start of a news program. This observation can help us to segment a video based on the program structures.

Then, in the news program itself, each subject is introduced by an anchorperson who could link the subject with another one. When a subject is introduced by an anchorperson, an icon most often appears at the top left or right of the screen. The icon contains a closed-caption and an image to represent the subject. Each subject is realized by a team which usually includes a cameraman, technicians, and directed by a journalist. Each subject contains a few shots (usually a dozen), together with a sound track. The sound track is composed of a short introduction by the journalist, then some speech segments spoken by other people, then a conclusion by the journalist. When somebody is interviewed by the journalist, his/her name and a very short description (place, title, etc.) appear at the bottom of the screen. At the end of a subject, the names of the people who have realized the subject may appear at the bottom of the screen.

5.1.2 On-Line CNN News Segmentation Using Scene Classifications

Based on the observation of general TV news production rules, we notice that a CNN Headline News program follows similar rules. It consists of distinct news stories that form small closed information units based on event structures with special editing effects in between, such as the appearance of an anchorperson, etc. For example, CNN news report video has characteristics of CNN news frames, such as the CNN logo, the model of the spatial structure of anchorperson shots, and the background when an anchorperson talks at the station. We may use these characteristics to differentiate CNN news video from others. In our experiment, we keep the CNN news structure as simple as we stated above. Therefore, by differentiating the anchorperson, we have identified the boundary of each story unit as shown in Figure 5.1.

Before anchorperson key frame classifications, a CNN video sequence is first split into different shots based on a particular criteria on either DCT domain features or visual features such as color histogram differences. Details of the scene change detection algorithms applied to RTP intra-H.261 compressed video stream were given by Zhou et al. ([108]) and discussed in detail in Chapter 2. We adopt the scene change detection algorithm in the compressed domain proposed by Yeo and Liu ([98]) for MPEG-1 video. A frame in the middle of the scene is selected as the key frame for every newly detected scene and used as scene indexing and summarization.

We performed an experiment of a binary classification of a CNN anchorperson to detect the news story boundary. The procedure is shown in Figure 5.2. We established a knowledge base on the CNN Headline News characteristics by training it using a 10-minute CNN Headline News video and extracting the histogram of the anchorperson key frame. For any incoming on-line key frame, we extract its features such as color histograms by using the same descriptor parameters as in our knowledge-base training. Features extraction agents might follow by clustering agents which can automatically cluster high-dimensional features into groups by unsupervised learning methods. The classification proxy agent checks if the new key frame's features are matched with classification rules of features stored in knowledge base and outputs a binary decision to indicate the concept category of the video, such as whether the key frame is CNN news or not, or if the key frame is CNN anchorperson or not. This is of great use in the sense that, once we can make this distinction, we can pass the video to a user if the user requires CNN news. We may also use the key frame of an anchorperson

Section 5.1. CNN News Video Segmentation and Indexing   **107**

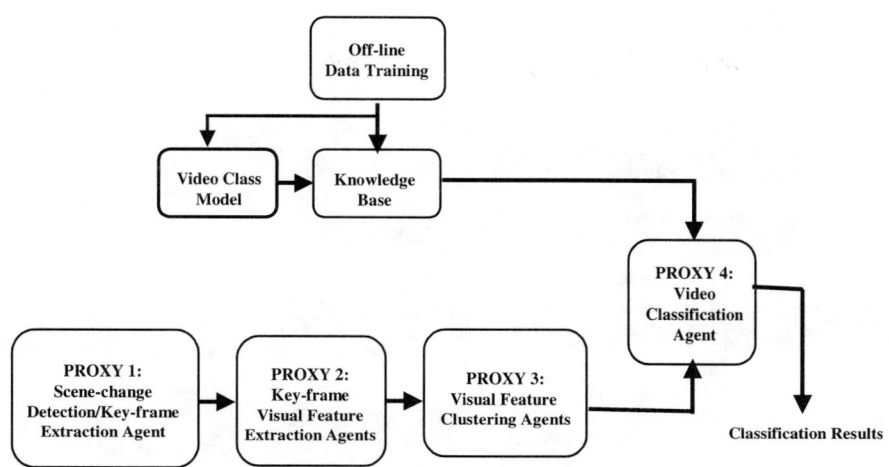

Figure 5.2. Agents for video key frame classification.

as the news story boundary for finer semantics extraction from the captioned text. By using the proposed algorithm to the classification of an anchorperson on a 30-minute-long CNN news program, we achieve an anchorperson classification accuracy of 93 percent on on-line real-time systems.

5.1.3 Table of Contents Generation for CNN News Indexing

After scene change detection and anchorperson detection, we can decompose the CNN news into units as shown in Figure 5.3. To generate the table of contents of each segmented story units in CNN news, we analyze the closed-caption text information according to visually detected story units (such as the boundary defined by an anchorperson in the CNN news) based on the above observed TV rules.

Different from general table of contents of news stories with key words (or a set of key words) to represent the semantic news content, we generate a hierarchical concept tree that allows hierarchical concept composition for retrieval and filtering. We used CNN Headline News during Nov.10 to 17, 1998, as the test data. An example of the CNN news concept tree is shown in Figure 5.4. We use key word matching and learned $TFIDF$ rules to map any new incoming video into one of the concept nodes of the hierarchy concept tree for indexing. In the filtering scenario, any new news data can be automatically classified by using either the key-word matching or the

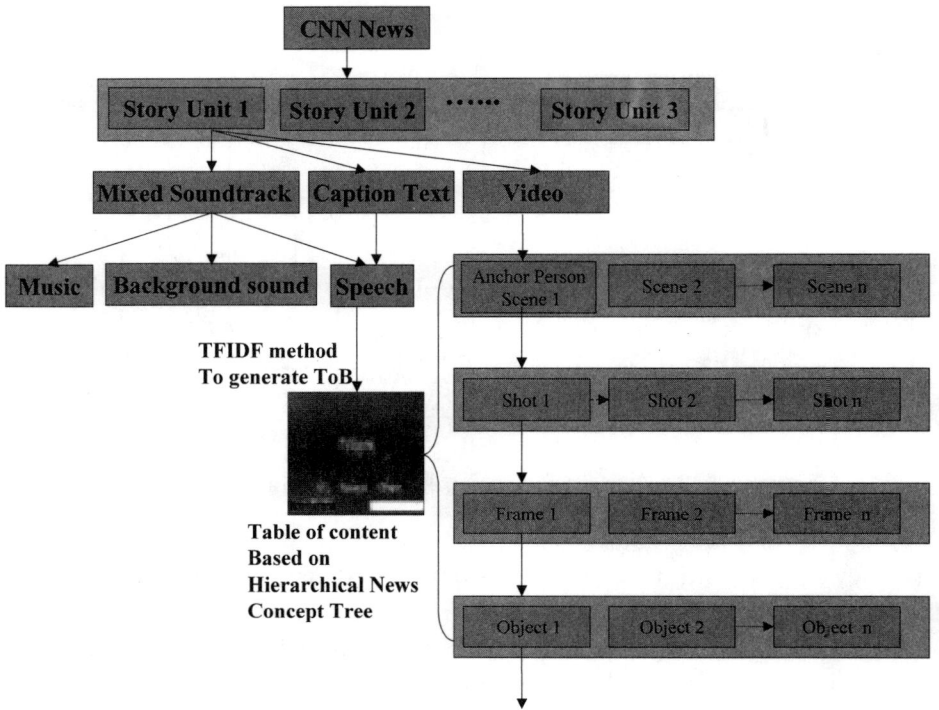

Figure 5.3. CNN segmented units.

$TFIDF$ classifier to see if the new data match the conceptual scope of users' requests. The table of contents of CNN Headline News video on Nov. 17, 1998 is shown in Figure 5.5. Once we get the correct boundary of story units, the average overall accuracy in concept classification for hierarchy concept tree nodes can reach more than 90 percent in our experiments on CNN news. Experiments of news filtering based on users' profile that follows the concepts produced in table of contents will be discussed in Chapter 6.

5.2 Hierarchical Video Classification on NBC 2000 Olympic Video

There are two types of video abstractions which should be segmented in sports videos. The first one is the segmentation of broadcasting programs.

Section 5.2. Hierarchical Video Classification on NBC 2000 Olympic Video

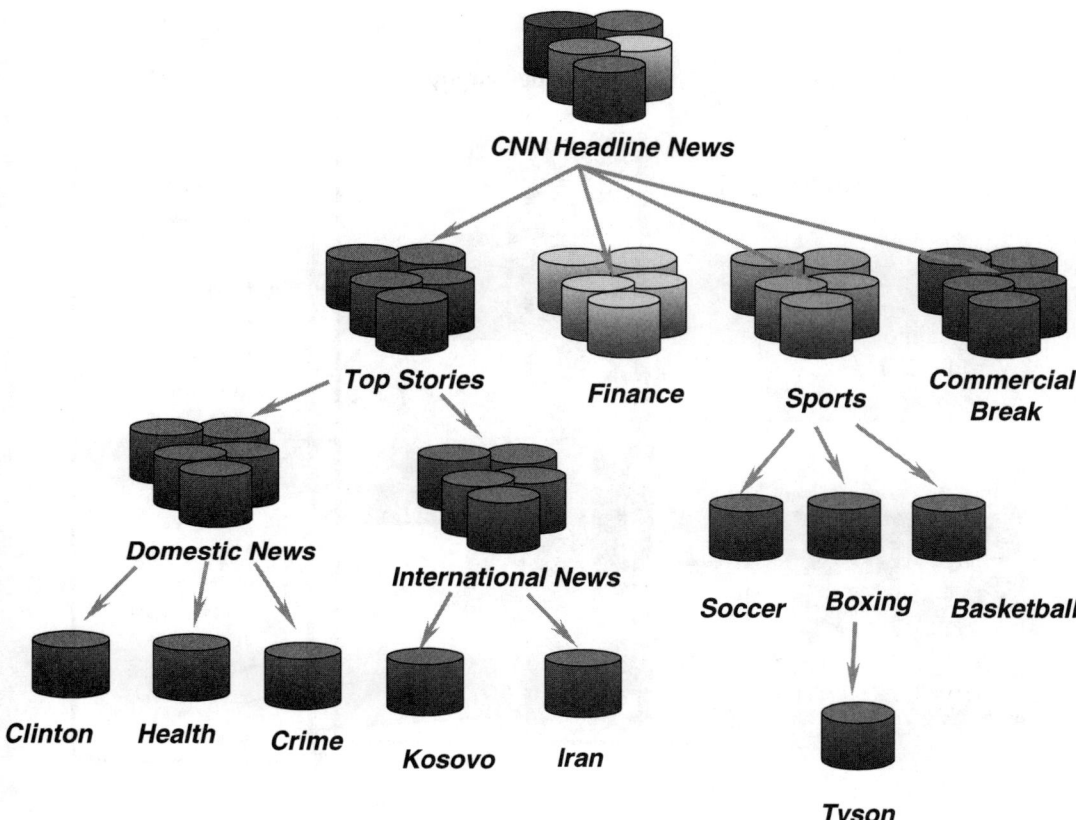

Figure 5.4. Video hiarchical concept tree for the table of contents of CNN news.

The second one is the segmentation of sports games into meaningful sports events. We recorded all broadcast programs covering the 2000 Olympic games by NBC, CNBC, and MSNBC and used them as experimental sports video data.

5.2.1 Program Segmentation of NBC Sports Video

Similar to a CNN news program, the NBC 2000 Olympic game program follows specific patterns too. Normally, the NBC 2000 Olympic game program starts with a station icon as shown in Figure 5.6(a). Then, an anchorperson [shown in Figure 5.6(b)] either summarizes the previous game or gives a brief introduction to the upcoming game and teams involved. The actual game

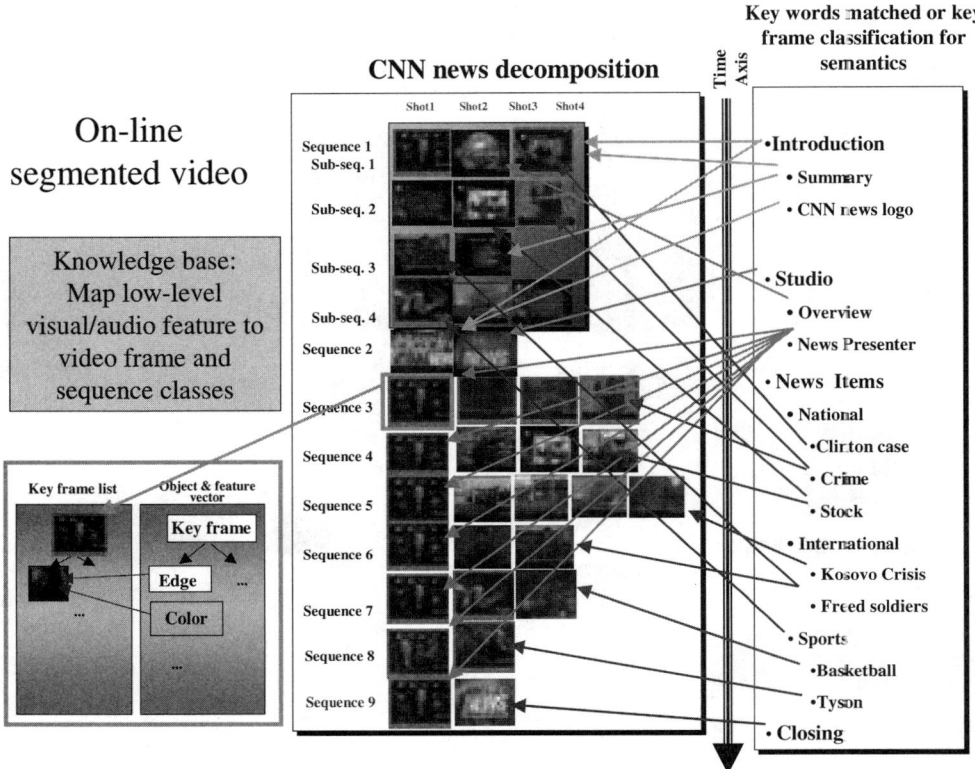

Figure 5.5. The table of contents of CNN news after segmentation.

starts right after that. Since commercial breaks are of little interest and relatively easy to detect using dark frames and high scene cut rate, etc., they are ignored in our analysis. We used techniques similar to those for CNN headline news analysis (Figure 5.2) to detect frames shown in Figure 5.6. Note that in our current implementation, we only used visual features to detect out program icon frames and anchorperson frames. A possible enhancement would be to use audio features to detect theme music. The fact is that there is Olympic theme music right before and after each commercial break, and the anchorperson generally appears right after the theme music. To detect a dynamically changed anchorperson, this theme music would help to detect a program boundary or an anchorperson more accurately.

Section 5.2. Hierarchical Video Classification on NBC 2000 Olympic Video 111

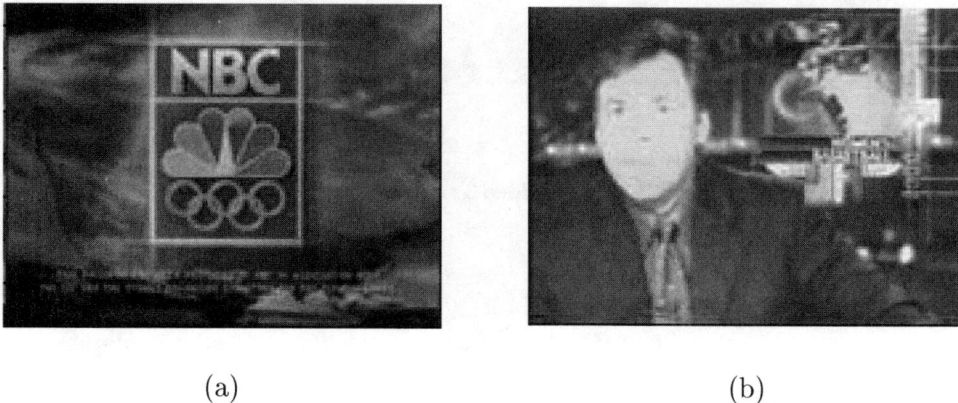

(a)   (b)

Figure 5.6. NBC 2000 Olympic game coverage: (a) program icon and (b) anchorperson

5.2.2 Sports Event Units Segmentation

The NBC sports event units segmentation procedure is shown in Figure 5.7. Video scene change detection is first applied to videos. A frame at the first 10 percent position of the scene is selected as the key frame for every newly detected scene and used as scene indexing and summarization as shown in Figure 5.8. Another heuristic rule used to segment the sports video is that a video event boundary is claimed if either a sound of a whistle or a new scene change is detected. Since a whistle sound almost always indicates a start or an end of an event at games, we treat it as a logical boundary break for a new event even if there is no scene change yet. Once the visual/audio scene boundaries and event boundaries are defined, we output the corresponding text based on the time stamp mapping with video/audio track. As closed-caption text shown on a TV screen normally has certain latency compared with true video scenes, we start to record the caption at the same time as the video begins, and stop caption recording at about 10 to 20 seconds after a video stop boundary is detected.

5.2.3 Low-Level Feature Extraction

Visual feature extraction is done for each key frame, and a clustering algorithm is used to classify key frames into their typical scene groups in each sports game. For example, we use dominant and regional colors and the edge information to classify a basketball key frame into one of four categories: the

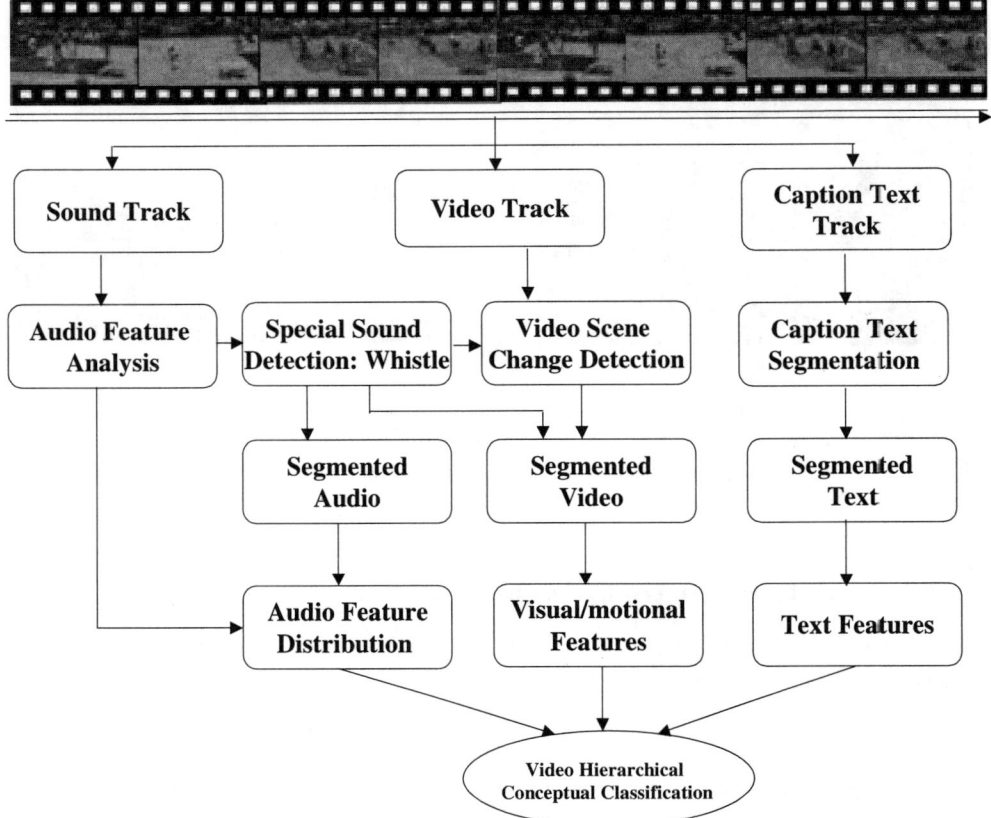

Figure 5.7. Sports event segmentation flow chart.

left court, the right court, the middle court, and others as shown in Figures 5.9(a) and 5.12. Figure 5.9(b) shows typical scenes which are often shown at the other four different games: water polo, softball, volleyball, and soccer.

In the experiment, the left court of a basketball game has a regional blue color in the middle of the left half region, and the edge around the region was horizontal and vertical with an angle. The middle field does not have the blue regional color. Instead, it has a circle edge in the middle. The percentage of each dominant color varies, too. Details of the key frame visual information analysis are described in our previous work ([107], [109]).

Among the low-level features that we utilized for our classification system are motion and compressed domain features, color features and edge features, acoustic features, and text vector features. Details of the features used in

Section 5.2. Hierarchical Video Classification on NBC 2000 Olympic Video

Figure 5.8. Scene summary of basketball video.

the NBC Olympic sports video classification are as follows.

5.2.3.1 Motion and Compressed Domain Feature Analysis and Extraction

Given that video is stored efficiently in a compressed format, the costly decompression can be avoided by analyzing the compressed representation directly. In addition, motion features are good cues to be used in video, as they are an integral part of a motion sequence. The macroblock motion information is typically calculated in MPEG encoders and is available in the compressed data stream. There are I-frames, P-frames, and B-frames in MPEG coded video as shown in Figure 5.10, where P- and B- frames are predicted by motion vectors with respect to a certain I- or P-frame. To quickly determine motion patterns for a certain video clip, we may focus on the direction and the magnitude of motion flows of P-frames only. P-frames provide forward motion prediction, which is useful in calculating the direction of the "motion flow" of the video clip of interest. Motion information is specifically important to sports video, where "motion flow" is an important indicator.

Motion Descriptors

Typically, we do not extract object motion because it is computationally complex and time-consuming. Instead, we try to use statistical motion descriptors since these descriptors give some macro information about the clip. We calculate statistical motion features for each video clip, including motion direction distribution, the dominant motion direction, and the average and standard deviation of the motion magnitude ([107]).

1. Direction of Motion Descriptor Extraction

For each motion vector in the vector image, we can cluster and then classify each motion vector's direction as shown in Figure 5.11. According to the motion descriptor defined above, we cluster the vectors based on the following criteria as shown in Figure 5.11(b): vectors in Region 1 and Region 8 are labeled as RIGHT; vectors in Region 2 and Region 3 as UP; vectors in Region 4 and Region 5 as LEFT; and vectors in Region 6 and Region 7 as DOWN. Then we calculate the amount of motion along each direction by counting the total numbers of vectors along that direction in each class for the whole video sequence. The percentage of motion along each direction is also calculated. This results in a four-dimensional motion direction vector. The most prominent direction in the direction histogram is the dominant direction of the video clip.

Section 5.2. Hierarchical Video Classification on NBC 2000 Olympic Video

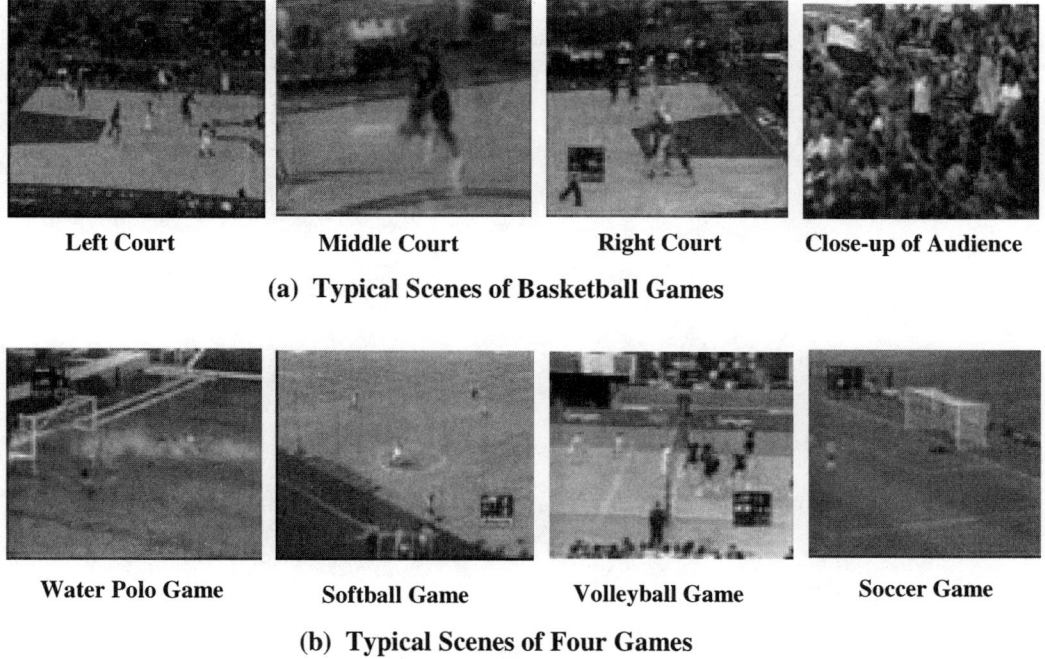

Figure 5.9. Examples of typical scenes of a basketball game (a) and water polo, softball, volleyball, and soccer games (b).

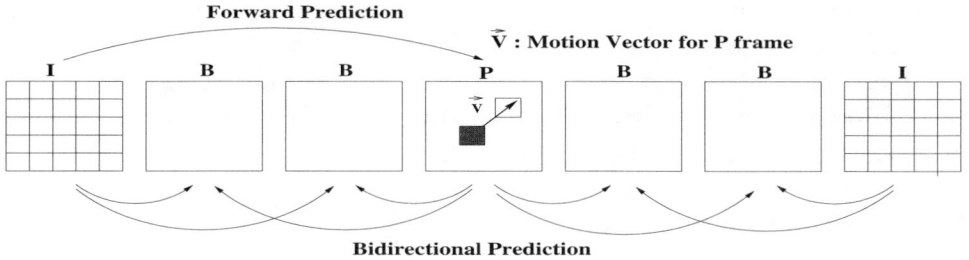

Figure 5.10. Motion compensation for forward and bidirectional prediction.

2. Motion Magnitude Calculation

Normally, the magnitude of the motion is also encoded in the motion vector. To get the instances of magnitude and speed of the motion descriptors along both the X and Y direction, we calculated the motion magnitude

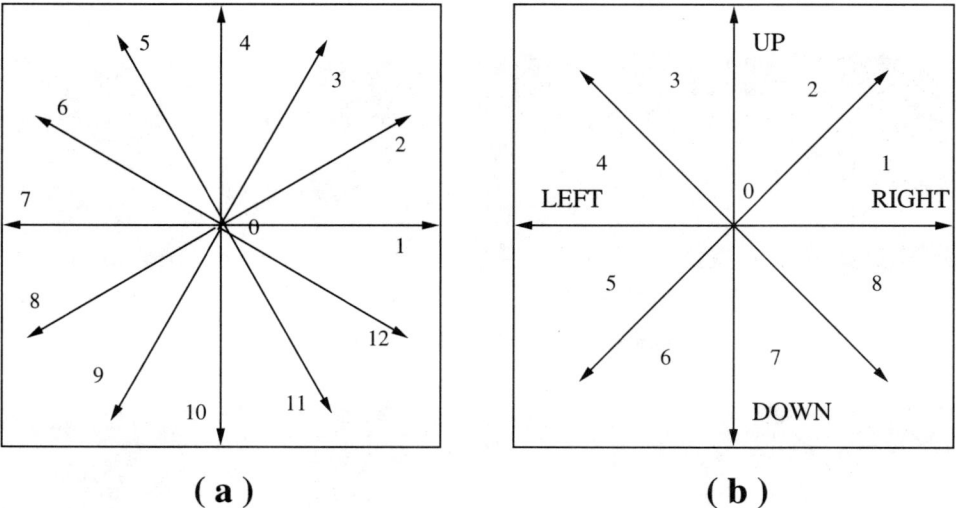

Figure 5.11. Motion directions extraction.

of the whole frame as Equation 5.2.1 and 5.2.2 :

$$x_{ave}(i,j) = \frac{\sum_{i=1}^{i=n}(x_i)}{n} \quad (5.2.1)$$

$$y_{ave}(i,j) = \frac{\sum_{i=1}^{i=n}(y_i)}{n} \quad (5.2.2)$$

where n and m are numbers of motion vectors in the frame with respect to the X direction and the Y direction, respectively. The maximum, the average, and the standard deviation of motion magnitudes along the X and the Y directions over the whole segmented video clip are then calculated and used as the attributes of the motion descriptor.

Compressed Domain Descriptors

Since I-frames are only intraframe coded, their bit rates only depend on the texture of the scene. For P- and B-frames, the generated bit rates are primarily motion dependent. However, texture also affects the bit rate due to the uncovered background of moving objects. Normally, for a low motion scene, macroblocks (MB) in P-frames and B-frames are mainly of the Forward predicted (FWD) type and the Bidirectionally predicted (BIP) type, respectively. As the motion increases, the forward prediction of MB in the P-frame may fail and, hence, the percentage of Intra (INT) MBs

Section 5.2. Hierarchical Video Classification on NBC 2000 Olympic Video

Figure 5.12. Various key frames of basketball scenes: (a) left court, (b) right court, (c) middle court, and (d) close-up images.

increases (nearly 100 percent in case of the scene cut). Similarly, in B-frames, the percentage of the BIP MB type decreases as motion increases. Based on the above observation, bit rates of I-, P-, and B-frames can be used as parameters in shot texture classification, while the percentage of Intra (INT), FWD, and BIP macroblock types in P-frames and B-frames can be used as parameters in shot motion classification. These descriptors were suggested by Dawood ([21]). The average and the standard deviation of distributions of each of above parameters [bit rates of I-, P-, and B-Frame, the macroblock percentage of Intra (INT), FWD, and BIP macroblock types in P- and B-frames] over the entire video clip are calculated.

5.2.3.2 Visual Feature Analysis and Extraction

In addition to motion, color and edge information also play an important role in object identification. In particular, we wanted to use color and edge information to classify a given video clip key frame into four categories such

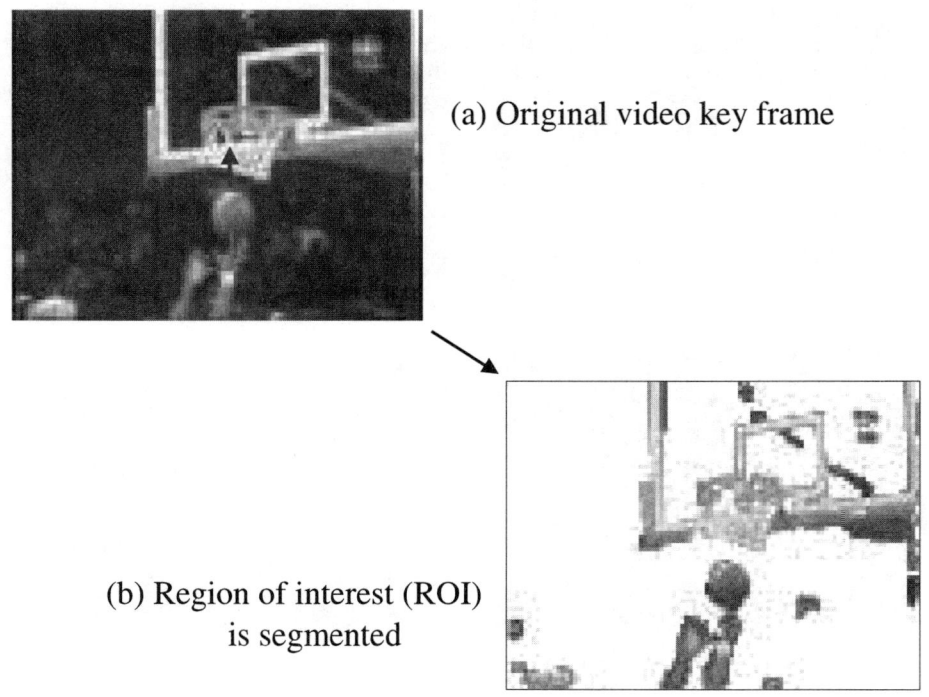

Figure 5.13. Regional color extraction for basketball video key frames.

as left court, right court, middle court, and others such as those close-ups shown in Figure 5.12. To do this, we extracted color information such as color histogram, dominant color, and regional color. YUV histograms are automatically generated by the shot change detection agents. To obtain regional color information and get more detailed color information and thus increase the differentiating power of the color feature, we considered the dominant color and the localized color information. We used the median-cut algorithm[35] to reduce the color map to about 256 colors. That is, colors in an image were mapped to their closest match in the new color map so that the colors of the original images were clustered. We used the clustering method to automatically detect a number of dominant colors and output them as a color tree for each frame. Then, pixels, whose distance to a dominant color was less than a given threshold, were back-mapped into a homogenous region with a dominant color feature. We got regions corresponding to the first five most dominant colors. The region centroid, the region boundary, and its area percentage over the whole image were also calculated as the attributes of the regional color for each homogenous region. Figure 5.13 shows the regional color extraction for one key frame which is possible for an event of scoring. We used a simple clustering algorithm to cluster features into different groups to reduce the dimensions of the features, and thus simplify the feature representation for further analysis.

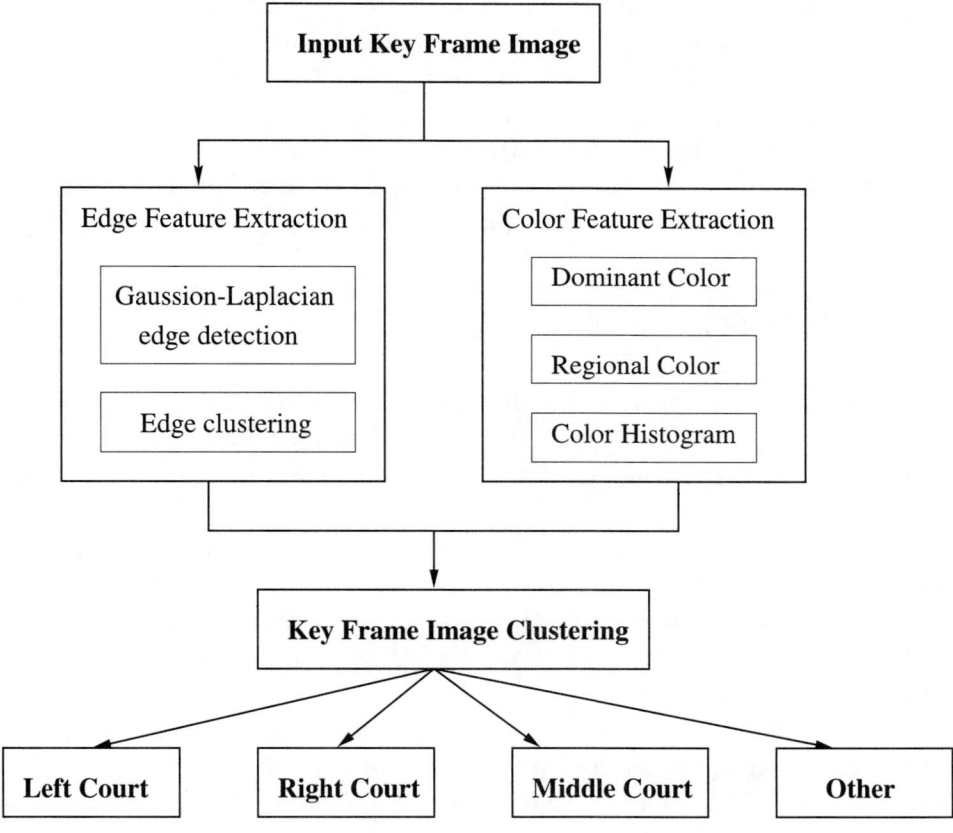

Figure 5.14. Feature extraction for key frame clustering.

Edge patterns can be used to analyze key frame characteristics and scene classification. For example, key images in a boxing match generally have fence-like edges which contain more lines of horizontal edges than those of soccer games. Edge patterns can also be treated as a simple texture in each key frame image. Here, we used a gradient edge operator to detect edges and took edge detection masks to fulfill the edge detection task. Edge detection masks included the first-order derivative masks, such as the Prewitt and the Robinson three-level masks. Edge detection with an option of the second-order Laplacian mask was also provided. In the first-order derivative masks, all eight directional masks, such as directions along horizontal, vertical, diagonal, and cross-diagonal, were used to detect edges along different directions. Edge densities along different directions and different lengths around the dominant color region were also calculated as edge features. Furthermore, we analyzed the edge information along four directions by distribution of visible edges and clustering detected edges into horizontal, vertical, and other category edge types.

Once we got color and edge features, they were automatically clustered into four key frame categories by unsupervised learning. The flow chart for key frame clustering is shown in Figure 5.14.

5.2.3.3 Audio Feature Extraction

To detect whistles in sports video and to find any useful features to differentiate sports or sports events, we extracted statistics of audio feature distributions along each segmented video sequence. The extracted audio features include *Short-time Audio Volume (SAV)*, *Zero-Crossing Rate (ZCR)*, *Frequency Centroid (FC)*, *Frequency Bandwidth (FBW)*, four *Sub-band Energy Ratio(SER)*, and *Spectrum Peaks(SP)*. Detailed description of the representations and analysis methods of these audio features are discussed in Section 3.5. A set of examples of the audio features from a basketball video is shown in Figure 5.15.

In sports video games, a whistle is a typical sound which often occurs right after a foul in basketball and soccer games, or at the beginning of a serve in volleyball, basketball, and soccer games. Whistles in sports video often last at least one second, and have stronger energy than speech and music. We implement whistle detection with both video semantic boundary detection and semantic meaning inference. The peak of the whistle spectrum normally ranges from 3500 Hz to 4500 Hz. We detect the most prominent frequency from the FFT transformed spectrum for every frame in an audio clip. It is claimed that the sound of a whistle is detected if there is a longer than one second window of peak frequencies which a fall into the range between 3500–4500 Hz, as shown at Figure 5.16.

Section 5.2. Hierarchical Video Classification on NBC 2000 Olympic Video 121

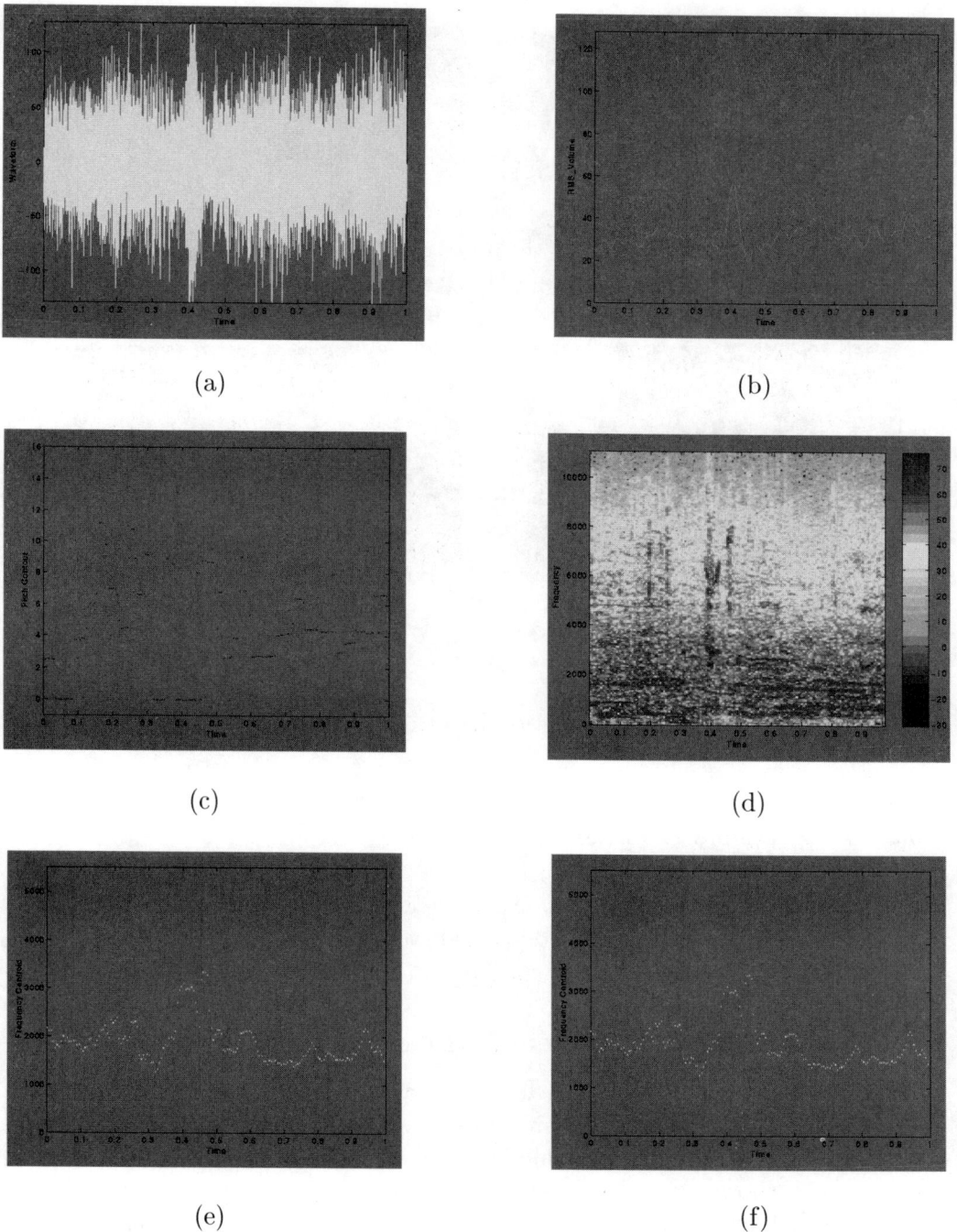

Figure 5.15. Audio features: (a) the waveform audio in basketball video, (b) the volume, (c) the pitch, (d) the spectrum, (e) the frequency centroid, and (f) the bandwidth of the audio.

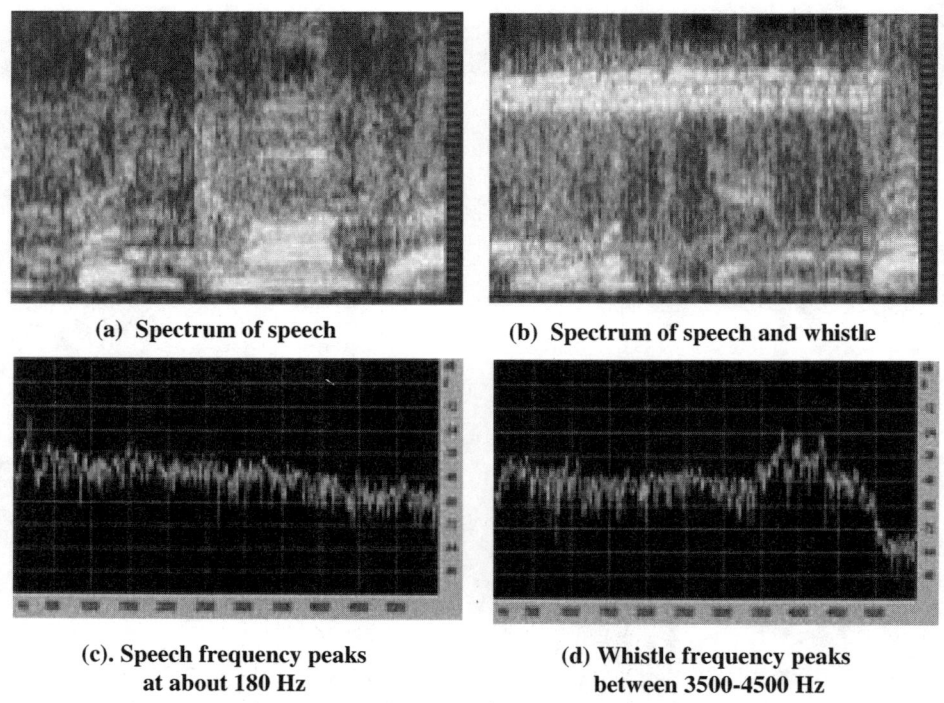

Figure 5.16. Whistle detection by using audio spectrum features.

5.2.3.4 Text Feature Extraction

Following the procedures of text feature generation and $TFIDF$ text classifier construction which were described in Section 4.5.3, we generate a $TFIDF$ vector for each node of concept in the concept hierarchy.

5.2.4 Rule-Based Video Event Classification for a Basketball Video

Before we go to hierarchical sports video classification, let's first study the rule-based basketball events classification based only on visual features to see how the knowledge-based and rule-based video classification system works. Figure 5.17 shows typical sequences in basketball game videos. Interesting basketball events include scoring, dunks, defense, offense, fastbreak, and fouls. It is very useful to index/filter video sequences based on these meaningful semantic concepts.

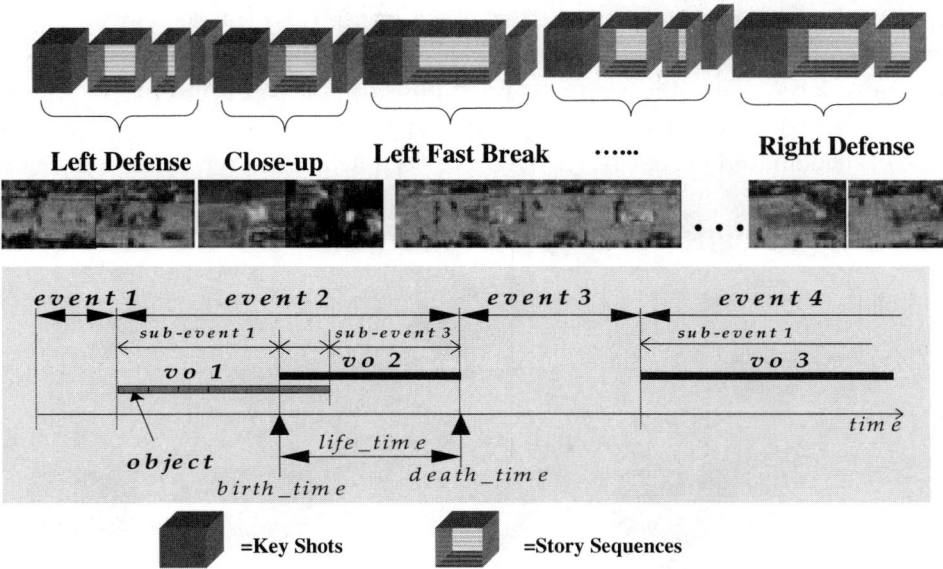

Figure 5.17. Basketball video sequences.

5.2.4.1 Decision Tree Training and Rules Extraction

We applied the proposed framework for classifier construction as shown in Figure 4.3 to a set of basketball video sequences for off-line training. Nine major events in basketball video were identified for video classification: 1. team offense at left court; 2. team offense at right court; 3. fastbreak to left; 4. fastbreak to right; 5. dunk in left court; 6. dunk in right court; 7. score in right court; 8. score in left court; 9. close-ups for audience or players. Sample video clips of different categories were identified and appropriate low-level features were selected. Then, we adopted an entropy-based inductive tree-learning algorithm ([75]) to establish the trained knowledge-base. This knowledge-base is represented by a decision tree in which each node represents an if-then rule as applied to a similarity metric utilizing an "appropriate" low-level feature along with a good "derived threshold." After training, we arrived at a three-level decision tree that contained 14 rules as shown in Figure 5.18. In the induced decision tree, a rule at each level is depicted as $<F, \theta>$. Note that the appropriate feature F and a good threshold θ are automatically generated from the training process. Note also that the semantic categories to be classified form the leaves of the tree. This decision tree provides a classifier. A new video clip is classified as follows. Following the tree, the feature utilized at Level 1 (i.e., the root level) is first extracted and the corresponding rule is applied, then a path is chosen. At the next level, the same step is carried out whereby an appropriate feature is selected and the corresponding rule is applied. Note that if a feature was already calculated earlier in the tree, we do not have to recalculate it again.

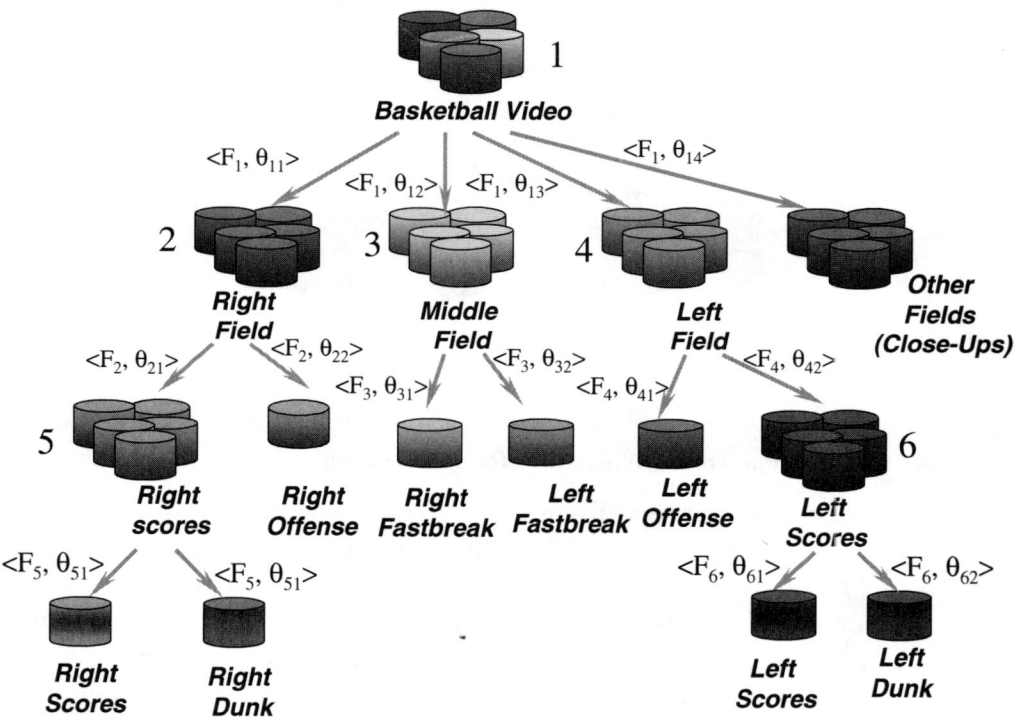

Figure 5.18. The rule-based tree for a basketball video.

From Figure 5.18, we notice that we only have to do at most three decisions to reach a single class in the classification stage. The number of decisions we arrived at is the same as the level of the tree. Also, no more than six features are needed to classify all nine basketball events. We used a set of basketball video data from one game to train the learning algorithm to get critical patterns and classifying rules that could differentiate

Table 5.1. Results of Basketball Video Classification by Using Both Motion and Visual Features

Class Name	Training Set sample #	Testing Set sample #	Testing Set Accuracy Percentage
Left Offense	20	14	78.5%
Right Offense	22	14	85.7%
Left Fastbreak	20	14	78.5%
Right Fastbreak	21	15	80.0%
Left Scores	15	12	75.0%
Right Scores	17	10	70.0%
Left Dunk	12	10	80.0%
Right Dunk	10	9	77.8%
Close-up Scenes	20	12	75.0%
Total/Average	157	110	78.2%

nine classes of events. The key frame type alone can be used to determine whether a video sequence is close-up or not. To make a decision on the right or the left fastbreak, we determine key frames, and then the average magnitude of motion vectors along the X direction for comparison with a learned threshold. In other words, we find that only the key frame type and the motion direction and magnitude along the X direction are relevant to right and left fastbreak classes. Thus, these relevant features are suitable for fastbreak event classification, indexing, and retrieval.

We applied the learned classification rules to classify a new set of basketball game videos. The experimental results are shown in Tables 5.1 and 5.2. Accuracy is defined as

$$Accuracy = \frac{\# \ of \ documents \ correctly \ categorized \ in \ class \ C_i}{\# \ of \ documents \ considered \ in \ class \ C_i}$$

We reach an accuracy of 70 to 85.7 percent for nine identified basketball classes as shown in Table 5.1 by including both motion and visual features in training and classification. For fast on-line video classification, we may trade-off the complexity of feature extraction and rule inference for the classification accuracy. We see from Table 5.2 that classification for the same data set with motion-only features gives an accuracy range of 64.3 percent to

Table 5.2. Results of Basketball Video Classification by Using Motion Features Only

Class Name	Training Set	Testing Set	
	sample #	sample #	Accuracy Percentage
Left Offense	20	14	71.4%
Right Offense	22	14	64.3%
Left Fastbreak	20	14	78.5%
Right Fastbreak	21	15	73.3%
Left Scores	15	12	66.7%
Right Scores	17	10	70.0%
Left Dunk	12	10	70.0%
Right Dunk	10	9	66.7%
Close-up Scenes	20	12	75.0%
Total/Average	157	110	70.9%

78.5 percent, which is reasonably good for fast on-line basketball classification and semantic category identification. The learned classification system is applied to the classification of basketball video into categories such as left fastbreak, right fastbreak, left dunk, right dunk, close-up shots, and so on. At a higher semantic level, if we know which teams are playing, such classifications can be used to answer queries such as "show shots where team A scored" as well as to support smart browsing of the basketball game.

5.2.5 Hierarchical NBC Video Classification

To evaluate the system described in Chapter 4, we recorded about 120 hours of Olympic sports video from NBC, CNBC, and MSNBC during the Sydney 2000 Olympic Games. Our experimental data included basketball, soccer, volleyball, water polo, and softball game videos. For each sport, we used video data of games played by one team as a training set, and then used another set of video data of the same type of game played by the same team against different teams as test data. The training data and test data for each sports game are from six game sets which are about each 45 minutes long. To evaluate the performance of classifiers at each node, we define the accuracy as

$$Accuracy = \frac{\# \ of \ documents \ correctly \ categorized \ in \ class \ C_i}{\# \ of \ documents \ considered \ in \ class \ C_i}.$$

Table 5.3. Classification Accuracy for Video Concepts at Level 2 (V/A Classifier: Video/Audio Classifier)

Class Name	Sample #		Accuracy	
	Training	Testing	Text classifier	V/A Classifier
Basketball	219	160	86.25%	88.9%
Volleyball	20	15	86.7%	86.7%
Soccer	20	20	85.0%	95.0%
Softball	20	15	80.0%	80.0%
Water polo	20	15	86.7%	93.3%
Total/Average	299	225	86.0%	89.0 %

Table 5.4. Classification Accuracy and Recall Rates for Video Concepts from Level 3-5 for Basketball Video (V/A Classifier: Video/Audio Classifier)

Level	Classes	Sample #		Accurarcy	
		Training	Testing	Text Classifier	V/A Classifier
Level 3	U.S. Team	199	145	80.7%	N/A
	Norway Team	20	15	80.0%	N/A
	U.S. Soccer Team	20	20	85.0%	N/A
Level 4	Active Games	157	110	64.5%	79.1%
	Game News	8	5	40.0%	60.0%
	Team Players	20	20	75.0%	30.0%
Level 5	Teresa Edwards	10	10	80.0%	N/A
	Lisa Leslie	10	10	70.0%	N/A

As shown in Table 5.3, both the text classifier and the video/audio classifier give good results for nodes at level 2. The text classifier gives an average

Table 5.5. Classification Accuracy and Recall Rates for Video Concepts at Level 5 for Basketball Video (V/A Classifier: Video/Audio Classifier)

Class Name	Sample #		Accuracy	
	Training	Testing	Text Classifier	V/A Classifier
Left Offense	20	14	64.3%	71.4%
Right Offense	22	14	57.1%	78.6%
Left Fastbreak	20	14	14.3%	71.4%
Right Fastbreak	21	15	20.0%	73.3%
Left Scores	15	12	58.3%	75.0%
Right Scores	17	10	50.0%	70.0%
Left Dunk	12	10	40.0%	80.0%
Right Dunk	10	9	44.4%	77.8%
Foul	22	15	93.3%	60.0%
Close-up Scenes	20	12	33.3%	66.7%
Soccer Foul	8	8	87.5%	50.0%
Soccer Goal	12	12	92.5%	41.7%

accuracy of 86.0 percent, the video/audio classifier gives an average accuracy of 89.0 percent. Although one classifier does not have a clear performance advantage over the other, independent agents could work independently and thus could improve the query processing speed based on either classifier for on-line media filtering.

For concept nodes on levels 3 to 5, the text classifier and the video/audio classifier perform quite differently. The results are shown in Table 5.4. The text classifier works very well in classifying the video/caption text content into U.S. teams with an average accuracy of 80.7 percent. It also gives very good query results on player's names, such as Teresa Edwards and Lisa Leslie of the U.S. women's basketball team in the 2000 Olympic games with an average accuracy of 75 percent. Note the performance of the text classifier could be affected by "typos" which are quite common in TV caption texts. However, it is very difficult to detect team's and player's names by using most visual/audio features, so we put "N/A" in the experimental results table. More sophisticated image/audio processing techniques, such as object tracking, face detection, or speech recognition, are needed to detect team players, which are often not available for on-line analysis. So the heuristic rules from the media cue optimizer will allow the text classifier to match with any new data if the query or filtering is related with concepts such as team name or player's name.

The bottom levels of sports video semantics are more related to visual/motional effects, such as fastbreak and dunk in a basketball game. Unless reporters specifically comment on these events, it is difficult to detect sports events purely from caption text streams. For very important game events, such as fouls in basketball and goals in soccer, it is easier to detect these with a text classifier, as reporters often comment on these events. So the text classifier gives us very good results for such heavily commented events but the video/audio classifier performs much worse. The video/audio classifier performs better than the text classifier on differentiating most less commented sports events, such as a fastbreak, and defense and scoring events in basketball games – with an accuracy rate from about 60 percent to 80 percent as shown in Table 5.5. These performance results are comparable to that of off-line basketball events detection. The classification performance gets worse in the hierarchy classifiers for Level 5 basketball event classifications than those shown in Tables 5.1 and 5.2. This is due to the errors passed from higher level concept classifications. The video/audio classifier also performs well on differentiating active games video versus game news video by giving an average accuracy of more than 71 percent for the Level-4 concepts. Because the test data set for game news is relatively small, to generate relatively large experimental errors (to increase one correctly classified data set the accuracy would increase 20 percent), the relatively poor performance with both types of classifiers for the "game news" concept may need more study based on a larger set of data to gain a better understanding of its true reason. Finally, we noticed that by choosing either text classifier or video/audio classifier, classification accuracies for all concepts nodes in the video semantic concept tree are in the range of 60 percent to almost 90 percent in our NBC 2000 Olympic game experiment.

5.3 Comparison with Previous Work

QBIC ([30]) uses low-level features, such as area and circularity in global shape features, to retrieve visual perceptually similar images and videos. However, due to the limited precision of these features and limitation to general applications, such an approach has limited expressiveness for answering queries with conceptual items and predicates.

In our approach, knowledge is structured and represented by using feature rules specified by the decision tree learning algorithm and $TFIDF$ at different levels of abstraction by the hierarchical concept tree and can be automatically generated. Our knowledge acquisition algorithm uses the techniques of supervised knowledge learning. In previous cooperative systems, no relaxation control facilities were provided for the concept relaxation nor the quality of the answer. Our system provides a mechanism to optimize the concept relaxation and includes the handling of feature relaxation control for video data querying. Classifiers at each concept node of superior performance are automatically chosen by using the media cues optimizer to achieve good classification, indexing, and query results.

Current research in multimedia database tackles the individual components of an overall multimedia database system. Models exist in representing images, sound, or video data ([7]). However, support of combined alphanumeric concepts and multimedia streams are required for more advanced applications such as on-line real-time multimedia streaming and filtering. Huang et al. ([38]) used the Hidden Markov Model to classify news video programs into commercial, basketball, football, news, and weather. The average accuracy was around 86 percent, and it was improved to 91.4 percent for the best performance with multiple classifier combination. However, they only conducted video classification at the first level of our video semantic concept tree. Besides, they used more complicated features than the simple features used in our on-line video stream analysis, and more complicated classification schemes such as combined classifiers. Our classification rules were simply concatenated together to train for the best rules of feature values for the most efficient video category partition, and the performance of our scheme is better than the counterpart reported by Huang et al. ([38]) with direct concatenation (the average accuracy is around 86.49 percent). Furthermore, our classification scheme is scalable in the sense of multigranularity.

Gong et al. ([32]) parsed TV soccer programs and claimed to achieve 80–90 percent accuracy on special key frame types, such as key frames of left penalty area, midfield, etc. Their performance on soccer events, such as top-left corner kicks and goals, have classification accuracy between 40 percent and 60 percent, much worse than the performance or our hierarchical semantic classifiers of the same level. Our on-line system can achieve much better concept classification results than their off-line system.

More recently, Park et al. ([69]) successfully used ontology to construct image concept hierarchy and the relationship between concepts. However, they demand human beings to be integrated into the learning loop by letting users specify the concept when they put any new images into the system. It is also not an on-line system.

5.4 Summary

We have implemented a scalable video classification system for video semantic classification based on concepts. The relationships of concepts are automatically embedded into the hierarchical concept tree. Flexible classification rules based on either visual/audio features or text features are monitored by the mixed-media cue optimizer which allows both fast and high quality video concept classification to support semantic-concept-based multimedia streaming and filtering.

An experimental study of the supervised rule-based video classification system was described in this chapter. Its application to basketball video was examined in detail. The rules were calculated by using an inductive decision tree learning approach applied to multiple low-level image features. There are advantages in utilizing such a rule-based approach. We applied the proposed classification system to basketball clips and achieved good results. In the experiment, we also tested the proposed on-line classification agents based on CNN News and NBC 2000 Olympic sports video programs, with special focus on concepts related to basketball game videos. Our experiments indicated that the semantic model as well as the low-level visual model were different in these two cases. (Figure 5.1 shows a typical semantic model for CNN news stories, while Figure 5.17 shows the story segmentation for a basketball game.) Within each semantic model, the corresponding low-level visual features are stored in the knowledge database for on-line feature matching.

Chapter 6

SYSTEM INTEGRATION

6.1 Introduction

Having described the major technical components for video content analysis in previous chapters, we are now ready to see how all these pieces fit together to form a complete system. Recall that we propose a system prototype for a content-aware and user preference-oriented multimedia data distribution system on the Internet based on the multicast protocol. Different from prior work on multimedia database systems which support primarily off-line data retrieval, our system will provide intelligent on-line feature extraction and semantic classification tools for video indexing, querying, and filtering by matching user's profile for real-time multimedia distribution and sharing over the Internet. This system is also expected to provide synchronized multimedia data stream distribution and filtering. In addition, our system attempts to organize multimedia resources on the Internet in a hierarchical and scalable way, allowing users to find items related to their interest based on the data content. The system can also be extended to applications such as interactive and personalized TV broadcast services, training services, and collaborative applications.

In this chapter, we focus on the use of the proposed infrastructure (as given in Figure 1.5) for automatic on-line multimedia content analysis and annotation to support semantic multicast system, and present a model of service that realizes the goal of providing effective on-line content-based media dissemination by filtering. More specifically, we develop a real-time intelligent system prototype for fast video content analysis and dissemination over the Internet.

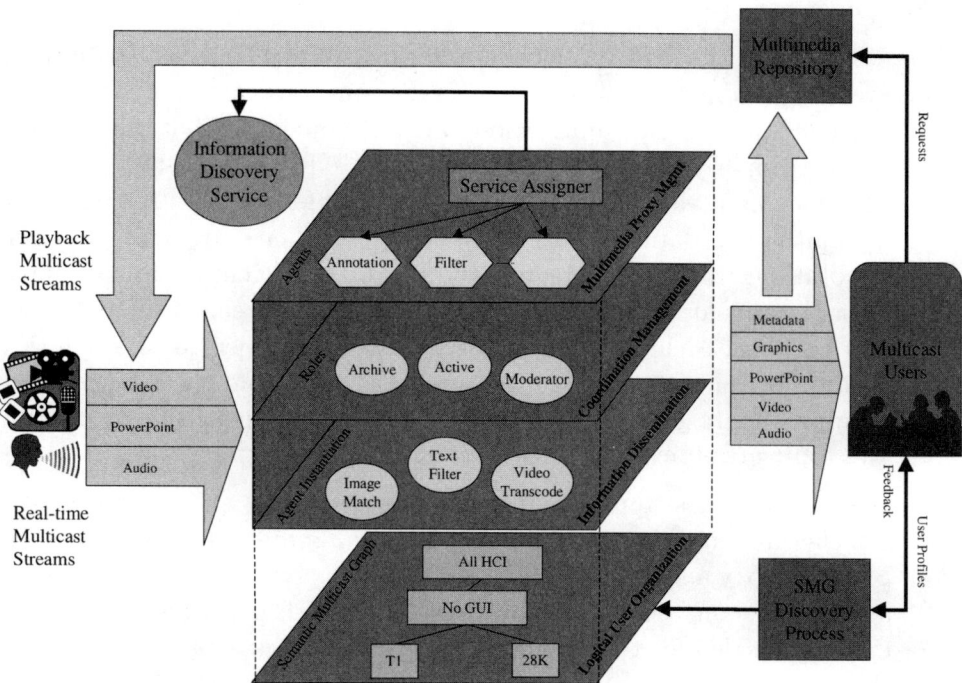

Figure 6.1. Semantic multicast system architecture.

6.2 Semantic Multicast System Architecture

Figure 6.1 shows the architecture of the semantic multicast system framework. The system contains the following modules:

1. Semantic multicast graph

   Given a collection of user requests against any real-time collaborative session of many overlapping streams, a semantic multicast graph is automatically constructed by a service assigner. A semantic multicast graph (SMG) organizes user requests into a hierarchy of group requests that represents an increasingly specialized range and quality of semantic topics as one goes down the hierarchy; it establishes a logical flow of collaboration stream between interest groups – how data streams should be annotated, manipulated, filtered, merged, archived, and propagated to effectively disseminate information to users belonging to different semantic interest groups.

2. Content agents

   To meet the goal of on-line information sharing and redistribution, every stream must be transmitted in real time or near real time. Quick content analysis and annotations are demanded so that the stream can be properly organized and indexed for off-line repository and on-line disseminated. Content agents gather information streams, such as multimedia data streams from the network or extracted/analyzed annotations from video streams, and incrementally enrich, archive, and filter streams for semantic coverage of nodes in the semantic multicast graph. Then, each information stream may be subjected to a "quick" on-line content extraction for feature descriptors and checking process to approximate general content semantics or descriptions (e.g., subject and topic areas) exhibited by the multimedia stream.

   In summary, most functionalities of content extraction, content analysis, checking and data filtering as per user interests are carried out at content proxy agents that are located in the networks. Here, the proxy agent is the generic concept of a machine which performs various operations needed for the on-line mode of data processing. For example, the annotation proxy agent generates annotations; the filtering proxy agent passes or blocks streams based on whether or not they fit a user's scope, and the transcoding proxy (and/or the summarization proxy) transforms the raw data so as to support low bandwidth or logically disconnected users. Note that it is possible to combine some of these operations into a single proxy. For example, the same proxy can perform annotations and filtering. The functionalities in content agents are the key research areas of this book has presented in Chapters 2, 3, and 4.

3. Service assigner

   The semantic multicast service assigner manages and coordinates the content agents. The semantic multicast service assigner and the collection of content agents it manages are collectively referred to as a semantic multicast proxy server.

4. Multimedia repository

   In this system, a multimedia repository stores and manages semi-structured multimedia data. Automatically extracted low-level features and semantic annotations of associated video by content agents are also organized in a formal schema and stored appropriately in a multimedia repository. These perceptual and conceptual annotations can be used as indexing and query mechanisms for video data management. An interface to a multimedia repository provides the tools for off-line data searching and viewing.

5. Information discovery graph

    The information discovery graph (IDG) organizes multimedia resources on the Internet in a scalable way, allowing users to find items related to their interest based on the content of the data. The IDG also provides a shared mechanism for announcing and finding real-time sessions in particular, and all multimedia data in general, that are available to view and/or join the collaboration sessions.

The contextual focus of a semantic multicast graph is realized by a hierarchy of content agents in the network that gather information streams and incrementally enrich, archive, and filter the streams for the semantic coverage of specific interest groups in the semantic multicast graph. In semantic multicast, a collaborative session becomes a multilevel multicast of data from sources through (multiple layers of) agents and to user interest groups. Each agent gathers the streams related to one or more semantic topic areas defined by nodes in the semantic multicast graph. Each information stream may then be subjected to an on-line content extraction process to approximate the general semantics (e.g., subject and topic areas) exhibited by the stream. Upon receiving an annotated information stream, an agent filters it into a level of detail appropriate for dissemination to its user groups and other agents subscribed to its output. Instead of filtering a single stream, an agent may instead merge two or more incoming information streams and disseminate the fused data product. In addition, agents may archive information streams and perform more detailed off-line analysis on the data to provide additional semantic structuring for subsequent retrieval.

The process that determines the optimal mapping of the semantic multicast graph onto a set of network content agents is referred to as semantic multicast agent coordination. During this process, the semantic multicast service assigner discovers, recruits, and coordinates the services of content agents to transform and/or enrich information streams to satisfy user groups' information needs. Through this mechanism, the semantic multicast framework amortizes processing on information streams by automatically moving common operations on information streams needed to satisfy multiple user requests to intermediate content agents providing the services.

6.3 On-Line Video Content Analysis Implementation

In this book, we propose and implement a prototype for on-line media content analysis infrastructure (as shown in Figure 1.5) to support user-oriented data dissemination over the Internet with the multicast protocol in Semantic Multicast project ([20]). In this infrastructure, each semantic multicast content proxy consists of modules to do one or more specific data processing operation(s) and runs as a daemon. Once the agent receives the raw data stream packets, it buffers, depacketizes, decompresses (if necessary), and processes them in a way dependent on the functionalities to be provided. The video content analysis agents are centered on a hierarchical decomposition of video data, and extract visual/audio/motional characteristic content by combining video semantic class inference engines. High-level video semantics are inferred from low-level features for the filtering purpose. Extracted features and semantic content can further serve as annotation or indexing for off-line database management. Some of the functions, which these proxy-based content agents perform, are:

- *Annotation.* The creation of a high-level annotation stream to tag an information stream is an important form of content enrichment and is essential to effective information dissemination in semantic multicast. An agent may generate tags on session substreams (e.g., timestamp or concept) to prepare for archiving and filtering. In its raw form, multimedia data types such as video and audio are not amenable to automated semantic interpretation and typically have to be enhanced with higher-level features such as key words, video scene change tags, and representative sample frames. For example, audio classifier can classify the audio signals into the categories, such as speech, noise, or whistle. And speech-understanding systems can automatically transcribe the audio stream in order to create a text of the spoken words which can be utilized to allow the creation of a time-aligned transcript of the spoken words contained in the audio stream. At the next level, natural language processing techniques can be applied to correct and summarize the transcript as well as to identify key words which will describe logical subunits of the entire session, as defined by the video segmentation operation.

- *Filtering.* An agent may select a session based on annotations to reduce the scope to the interests of a particular group. Such filtering is generally time-constrained to minimize the latency incurred in the delivery of filtered information to users. In our filtering algorithms, we come with a knowledge base which can accommodate the media low-level feature descriptor plus description schemes to facilitate the filtering. Each feature descriptor has its own specific definition and extraction operator. Filter agents make data redistribution decisions based on the extracted feature values and stored rules in the knowledge base.

- *Archival.* An agent may store "appropriate" subsessions in an associated multimedia archive. As the agent archives the stream, it performs a more detailed and off-line analysis to provide additional semantic structuring and indexing for subsequent retrieval and feedback to the semantic multicast graph.

- *Temporal synchronization of content with descriptions.* To allow the temporal association of descriptions with content (AV objects) that can vary over time and effective media stream consumption, we use timestamps of RTP media streams as a synchronization connection between various media streams.

- *Synthesis of multiple low-level features associated with a content item.* An agent will allow flexible localization of descriptor data with one or more content objects. A variety of descriptors and description schemes could be associated with each content item. Depending on the user's profile, not all of these will be used in all cases. In push applications (e.g., real-time multicast/broadcast), effective feature synthesis and data multiplexing are needed to satisfy various content requests from users.

- *Buffer management for further decisions and actions.* Local storage and buffering management is essential for real-time applications, as every step of content analysis or decision-making generates latency.

6.3.1 On-Line Scene Change Detection and Key Frame Extraction

Video is a rich multimodel source; it contains audio, moving image sequences, and text information. These different media display distinct characteristics, and express information at different levels and in different degree of detail. While trying to annotate and extract semantics from such data, it is necessary to utilize information available from all sources so as to improve the quality of the created metadata. In our system implementation and evaluation, audio, video, and caption text are recorded and demultiplexed at the same time from the broadcast CNN news by using one of the RTP tools[1] – *rtpdump*; the recorded video, audio, and caption text are multicast to separate multicast channels later as experimental collaboration data by using another RTP tool – *rtpplay*.

Scene change detection is very important, since it is the first step and most fundamental element for further video analysis. We have developed a content agent that implements a scheme that summarizes real-time video in terms of a few representative key frames from video sequences by the scene change detection process. The processing scheme is shown in Figure 6.2. The main component of this scheme is based on real-time scene change detection on RTP intra-H.261 compressed video streams such as that commonly used in MBone broadcasts. A video is split into meaningful scenes such that each frame image that corresponds to a different shot is detected out as a key one as based on a particular criterion. Figure 6.2 shows a scene change detection based on histogram comparisons between adjacent frames of the video stream. A video clip can then be summarized via key frames extracted from different scenes that make up the whole stream. We take into account changes in both luminance and chrominance values in making the decision. If the change in the luminance or chrominance histogram over successive frames is larger than the threshold, we categorize that frame as a scene-change frame. A key frame is generated from that stream and sent out as a scene in a multicast channel. To avoid unclear images caused by editing effects such as dissolve, we choose the frame that sits at the end of the first 10 percent of video sequences after a new scene has been detected, as the key frame of the scene. The scene-change tag and key frame can also be put into a multicast channel so that other agents that are attempting to do other kinds of processing on the network can utilize this extracted information.

Chapter 2 described a family of algorithms, including one that utilizes full decompression to get full frames and a partial decompression to get information on the changed blocks so as to estimate the extent to which the full frame has changed since the last frame. In earlier investigations, joint algorithms based on video codec characteristics were carried out to acquire

---

[1] www.cs.columbia.edu/~hgs/rtp/

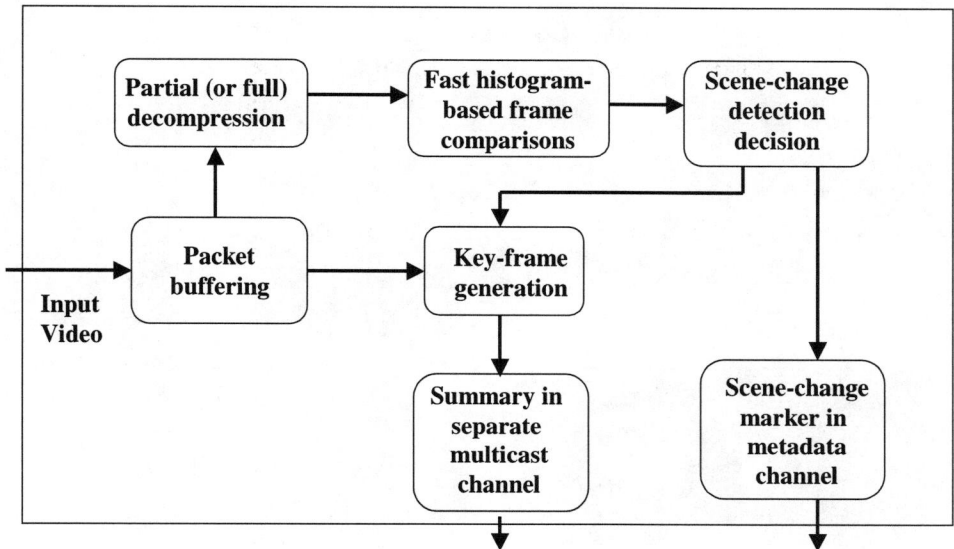

Figure 6.2. On-line scene change detection for H.261 video.

fast and accurate scene detection. Experimental results showed that our algorithms are capable of supporting real-time video processing and satisfying on-line annotation needs ([108]) given different network and data characteristics.

6.3.2 Low-Level Feature Analysis and Extraction

Video low-level features are widely used as index mechanisms for multimedia databases. Furthermore, patterns shown by the low-level features of different types of video can classify video into semantic categories. Thus low-level feature analysis and extraction are important in video annotations because they can help indexing and filtering the on-line multimedia. All features implemented by content agents are shown in Figure 6.3.

6.3.2.1 Visual Feature Analysis and Extraction

The implemented features of moving image sequences include motion features, color features, and edge features. Note that the set of low-level features could be more complete to comply with the MPEG-7 standard and they could be varied for different applications.

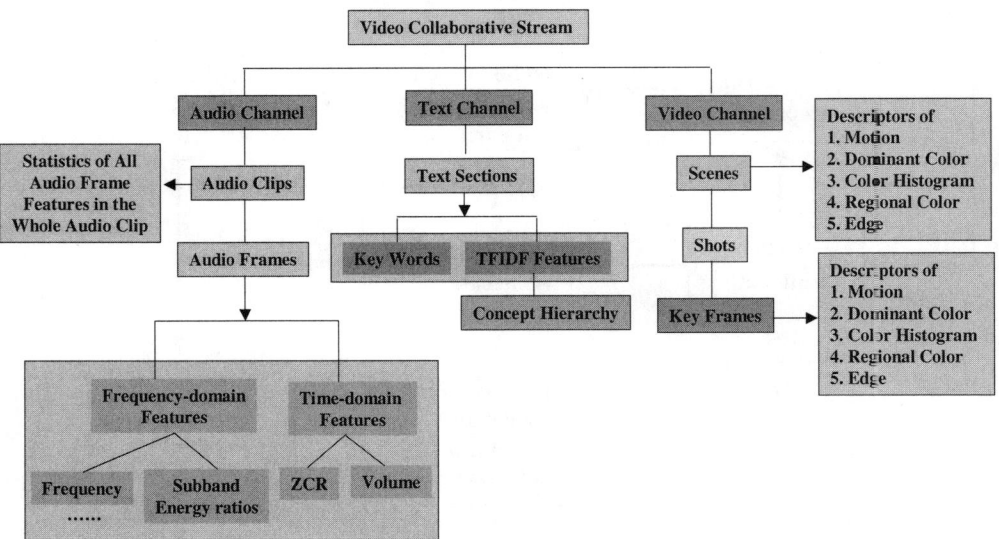

Figure 6.3. On-line extracted features of H.261 video, audio, and caption text streams.

We extract the color histograms, dominant color, regional color, and centroids of the dominant and regional colors of key frames of a scene. For some of the low-level feature extractions, we also use the unsupervised method to cluster features into different groups, thus reducing the dimensions of the features and simplifying the feature representation for further analysis. One example is the edge information where we cluster the edge patterns of each frame into the horizontal, vertical, slant, regular, or nonregular categories.

6.3.2.2 Motion Feature Analysis and Extraction

A video sequence consists of several background and foreground video objects. Precise descriptions of the motion in a video sequence is possible by describing the motion of the background and all foreground objects. The background motion is generally related to that of the camera. Descriptions of motion of foreground objects are generally more complex. Motion information of both background and foreground are very important for video content analysis. Because vic tools use the intra-H.261 codec ([54]) which lack a motion vector for both camera motion/editing and object motion information, we described the object motions by using an approximate measurement by considering the codec characteristics of intra-H.261. As we stated in scene change detection agent section, we have changed macroblock information. We use the translation of the macroblock centers between the two consecutive frames as the frame motion features. At the same time, bit-rate is also a good indicator of changed macroblocks in intra-H.261 codec. A large frame bit-rate indicates that many intra-coded macroblocks are encoded and a highly moved objects may exist in the frame. In news videos, anchorperson frames generally have less motion, thus their bit-rates are smaller compared with those of news stories.

For MPEG-1 video, we calculate motion direction distributions of the video sequence corresponding to each scene along a few directions such as up, down, right, and left. We also calculate the average and the standard deviation of the motion magnitude of the video sequence. We extract the color histogram, dominant color, regional color, and centroids of the dominant and regional colors of key frames of a scene. For some of the low-level feature extraction, we also use the unsupervised method to cluster features into different groups, thus reducing the dimensions of features and simplifying the feature representation for further analysis. One example is the edge information, where we cluster edge patterns of each frame into the horizontal, vertical, slant, regular, or nonregular categories.

6.3.2.3 Audio Feature Analysis and Extraction

We implemented two categories of audio features: time-domain and frequency-domain features. The details of the audio feature representation and extraction algorithms were described in Chapter 3. Here, we mainly implemented audio feature analysis agents for special audio events detection, such as whistles in sports videos.

6.3.2.4 Text Feature Analysis and Extraction

We implement a simple key word-matching algorithm for video semantic understanding and filtering. At the same time, we adopt a $TFIDF$ algorithm ([51], [78]) to classify caption texts which correspond to the interesting video sequences and audio clips into a hierarchical semantic category tree.

6.3.3 Knowledge-Based Video Classification

In order to classify incoming video streams into meaningful semantic classes, we should classify video with the smallest number of features that can be easily extracted. The proposed rule-based video classification system (shown in Figure 4.3) in Chapter 4 is good for both on-line and off-line video classifications, which are applicable to video indexing systems, video scene understanding and mining, on-line video filtering and video intelligent summarization, fast video browsing, and so on. A general video classification problem can easily follow the system prototype illustrated in Figure 4.3. It is even more straightforward to apply such a rule-based video classification system to on-line user-profile specified video filtering. Figure 6.4 illustrates the data flow for such applications. For any user-specified video category, the knowledge base contains the corresponding characteristic features and rules to identify it. Only those relevant features are extracted on-line and they are matched with the rule threshold. If it satisfied the rule, then the real-time stream matches the user's expectation. Otherwise, it is not, and further decision based on this intelligent video classification would be made based on the application. On-line knowledge-based video classification agents are implemented to fulfill the video semantic conceptual content analysis functions.

6.4 Content Agent Coordination

In the current design of semantic multicast, the negotiation of agent services is realized by requiring each proxy to implement an "applicability" function that programmatically captures the behavior and capability of the agent and exposes it to the semantic multicast framework. However, an agent often can process the input data streams but not necessarily transform them into a form that satisfies the target group request. In this case, while the agent cannot by itself completely service the data processing needs, it may still "partially" transform the input data streams into others that can be further processed by other agents to satisfy the target group request. As a result, the applicability function is defined to return (i) an intermediate group request that the agent can process the input data streams into, (ii) the specification of the intermediate data streams, and similar to the last case (iii) the configuration parameters for the agent to perform the operation. The intermediate group request and data streams can then be treated as a new source request and data stream and, along with the target group request, get passed to other agents to determine if they can complete the mapping from the link between the source and target group requests to a series of agent instances.

Section 6.4. Content Agent Coordination   **143**

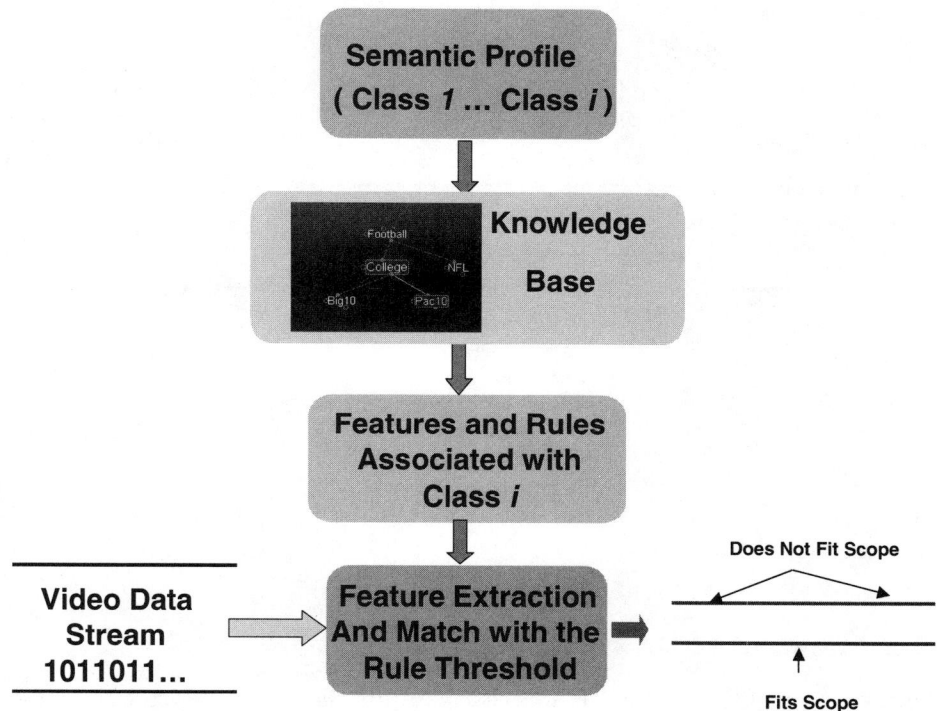

Figure 6.4. On-line knowledge-based and rule-based video classification for video filtering.

6.4.1 Communication Between the Proxy and the Service Assigner

This section describes the parameters that are passed between the proxy and the service assigner. As shown in Figure 6.5, the proxy lets the service assigner know of its capabilities in terms of the functionalities it offers (e.g., key word generation, scene change detection, etc.), the parameters it has to be supplied with (e.g., minimum expected degree of accuracy, some kind of threshold level as an input to the algorithms etc.), and its current load, whether it is free or busy, etc. The service assigner is then able to distribute new content analysis tasks accordingly.

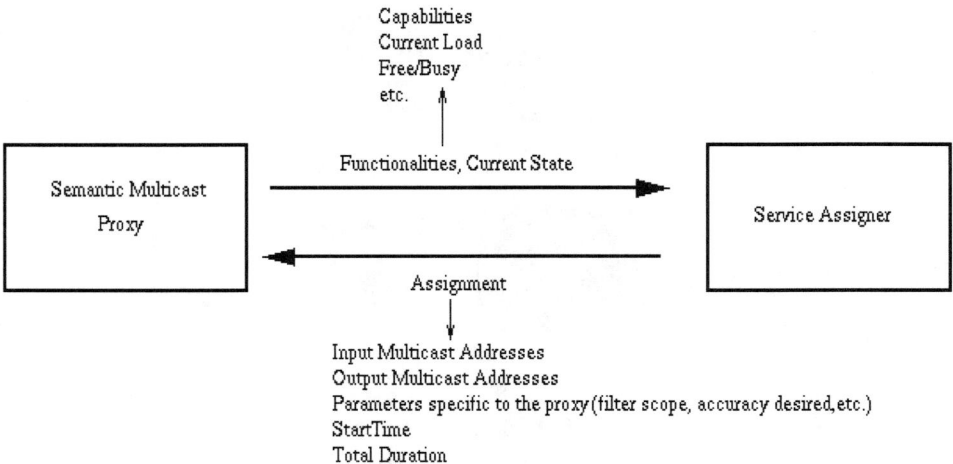

Figure 6.5. Communication parameters between the service assigner and the semantic multicast proxy.

6.4.2 Concept of the Metadata Channel

There are currently several IP multicast applications which are being used for collaboration – vic([54]) for video, vat[2] for audio, mb and wb[3] for whiteboard, and so on. (For an extensive list of several multicast applications that are publicly available see http://nic.merit.edu/net-research/mbone.) Note that these are simply raw data streams that are encoded and transmitted at the source site and decoded at the client site. These raw data streams by themselves are not conducive for making any kind of sophisticated decisions at the semantic level, such as whether the current data fits a particular scope as defined in a filter. Thus, we need to process raw data to arrive at new representations that will enable the operations stated above. We call these new representations metadata or annotations.

As media content analysis agents continue to generate media annotation, which would be various low-level feature extractions or concept classification results, these results should be sent out for any user to search for media from different channels. At the same time, these annotations should also be sent to a multimedia repository to facilitate off-line multimedia retrieval. Fur-

---

[2] http://www-nrg.ee.lbl.gov/vat/
[3] http://www-nrg.ee.lbl.gov/wb/

thermore, content agents many times need to analyze features based on previously extracted annotations other than original raw data. This brings up questions such as, "Where should these metadata be carried?" and "What format should be used?", etc. This is an especially important issue with respect to the current MBone since the raw data such as video, audio, and wb are carried in separate channels and not multiplexed (a design choice to let viewers choose only what they want to receive). This is unlike other standards such as MPEG-2 or MPEG-4 where the video, audio, and private data can be multiplexed together to a single channel. In addition, there is also the issue as to whether the metadata channel should be carried using reliable multicast protocols instead of unreliable protocols, and there are two possible methods in this regard. One is to carry the metadata in the same channel as the raw data. In this case, the metadata which is generated using a particular type of raw data is attached to the same data stream. One of the possible advantages of this method might be that it is easier to synchronize the metadata to the raw data. Such a method also has several disadvantages:

- RTP protocol has separate packetization formats for different video and audio compression codecs. Thus, an annotation proxy will have to closely understand the packetization formats for all these individual codecs. In addition, these packetization formats are standardized and constrains the choices for the annotation payloads.

- RTP protocol is lossy. Thus, the metadata which are transmitted along with raw data which are carried in the RTP protocol can lose the metadata along the way. In some cases, the metadata might not be useful once the actual data has been lost. However, these could be scenarios where the end user or a filter proxy wants to get the metadata effectively even if the raw data is being lost.

The second method is to have a separate metadata channel carry annotations related to all data types. The advantages of this approach are as follows:

- The metadata channel can be carried using the reliable multicast protocol, which might be better suited for the nature of the metadata rather than unreliable protocols.

- Some metadata could be generated using a combination of two different types of raw data. An example could be one where both video and

```
Metadata Packet Format
```

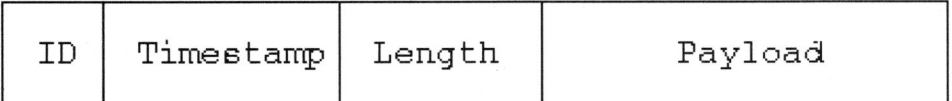

Figure 6.6. The format of metadata packets.

audio are used together to make a concept identifier decision. In such cases, we will have to attach the metadata to both streams for the other approach. In this case, it would be better to have just one copy of the annotation in the common metadata channel.

- Finally, we have more flexibility with respect to the formats. (We can also contribute to other standards such as MPEG-7, which attempts to provide similar functionalities.)

The one possible disadvantage of this method might be that it is more difficult to synchronize the metadata with the raw data. The current design decision is to use the latter approach of having a separate metadata channel. The current design decision is to use a separate metadata channel concept and tagging techniques to transport and integrate the content annotation as additional streams accompanying the raw audio, text, or video streams. Users can select information more appropriate to their uses and desires from a multicast stream of many channels, while using the same meta-information from the metadata channel as that used in the search.

6.4.2.1 Metadata Packet Format

The metadata channel is designed as the sole channel carrying metadata generated or associated with various sources. We defined a fixed metadata packet format as shown in Figure 6.6. It identifies the type of metadata contained in a packet as well as the format of the payload as follows:

- Metadata ID – 8 bits: This is the byte that identifies the type of metadata. It is possible to have a two-stage ID, the first stage identifying the raw data stream to which each metadata is associated with, and the second stage specifying the exact type of metadata. In our experiment with CNN news video analysis, we kept the metadata format

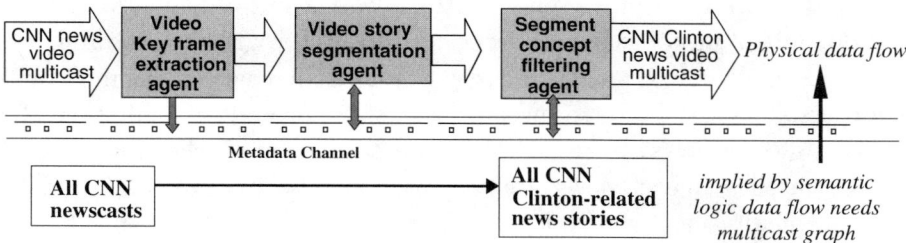

Figure 6.7. Agents cooperating to realize a semantic multicast graph.

simple and it had only a single stage ID. We generated four types of metadata: key frame tag, color histogram, key frame classification, and caption text, and granted each type of metadata a unique ID.

- Timestamp – 32 bits: The timestamp denotes the sampling instant of the metadata or the start duration from which the metadata is relevant. For some types of metadata there should also be a specification of the duration for which the metadata is relevant. This is not true for all types of metadata. The duration should be included as part of the payload. The timestamp will also allow synchronization of the metadata packet with the raw data stream.

- Length of the Packet – 32 bits: This specifies the total length of the packet including the 9 bytes that denote the header part of the packet.

- Payload: This is the part that contains the actual metadata. The payload format will vary with the type of metadata. Each type of metadata should have its own payload format. A more sophisticated metadata packet format should contain a field that identifies the proxy, version numbers, extensions, etc.

6.4.2.2 Video Analysis by Coordinating Agents

Figure 6.7 shows how data streams carrying "all CNN newscasts" may be processed by a series of semantic multicast content agents to produce data streams carrying "CNN Clinton-related stories" that satisfy the request represented by the corresponding semantic multicast graph node. In particular, a video scene change detection and key frame extraction agent may identify natural breakpoints in the newscast video and a video segmentation agent may use that information to separate the newscast into segments representing independent news stories depending on the program structures. For example, in CNN news, a natural story boundary normally is defined by the anchorperson as he/she introduces a news story or ends an old story. Furthermore, after a newscast program has been segmented, the closed caption text associated with each news story segment may be analyzed and processed by a concept-filtering agent to identify news stories related to a specific topic for distribution.

There are three proxy agents shown in Figure 6.7: the annotation proxy, the segmentation proxy, and the filtering proxy. The inputs and outputs for each proxy are specified below.

- Inputs to the annotation proxy.
  Multicast channels containing raw data or other (typically lower-layer) annotations that have been generated elsewhere are carried in the metadata channel. In this experiment inputs to a key frame extraction agent are the raw data of intra-H.261 video.

- Output of the annotation proxy.
  Annotations are generated to become the output onto the metadata channel. In our experiment the output of the key frame extraction agent was the color histogram of the key frame and the raw data of the key frame image in decompressed format.

- Inputs to the video segmentation proxy.
  Video segmentation needs concept classification of the video key frames or sequences. The on-line CNN news story segmentation is realized under the semantic multicast framework by automatically combining agents that implement on-line video scene-change detection, video key frame extraction, low-level feature extraction, feature clustering, and video classification, as shown in Figure 5.2. Thus, inputs to the video story segmentation proxy are outputs of the agents mentioned above. To classify on-line anchorperson key frames quickly, we establish a knowledge base (See Figure 5.2), which contains rules to classify video by mapping the characteristic perceptual features with that of CNN news anchorperson images. Once an anchorperson is detected, the story unit boundary is also determined. A table of contents of CNN news videos (See Figure 5.5) is then generated by analyzing captions

Section 6.4. Content Agent Coordination 149

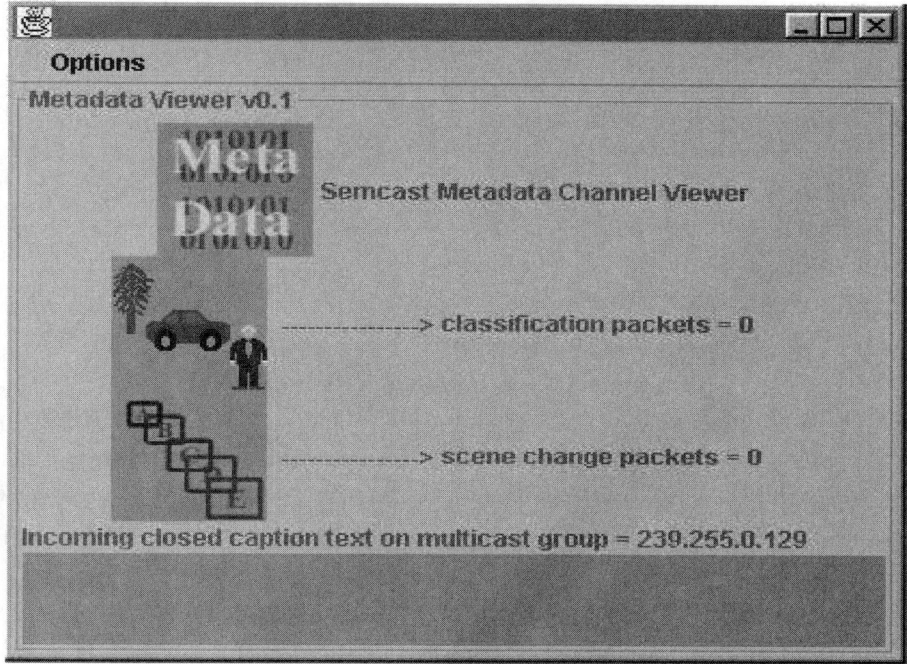

Figure 6.8. The Metadata Viewer (MV) shows the metadata associated with a certain video stream.

in each segmented video sequence. The Metadata Viewer (MV) shows all generated metadata in Figure 6.8.

- Output of the video segmentation proxy.
  For any incoming on-line key frame, video feature extraction agents extract low-level features quickly by using the same feature descriptors and algorithms as those specified by the knowledge base. A key frame classification proxy checks if a new key frame's features are matched with those in the knowledge base, and a binary decision (Yes/No) for each key frame semantic category results. Only when the result of the key frame is matched with the anchorperson is the story segmentation tag attached/sent. This is of great use in the sense that once we can make this distinction, we can either pass the video to the user if the user requires CNN news, or we can use the key frame of the anchorperson as the news story boundary for finer semantic extraction as from captioned text, which makes it possible to generate a table of

contents of CNN news (as shown Figure 5.5).

- Inputs to the filtering proxy.
  Inputs to a filtering proxy are (1) raw data of all video, and (2) audio and caption text, together with the metadata, such as, key frame classification tag of the video and its timestamp from the metadata channel and the user's scope.

- Outputs of the filtering proxy.
  Transformed raw data in new multicast channels (i.e., blocked, passed through media, or summarized video) are outputs from a filtering proxy. In our experiment, only video sequences containing stories which were related to President Clinton were distributed to designated users who made the request.

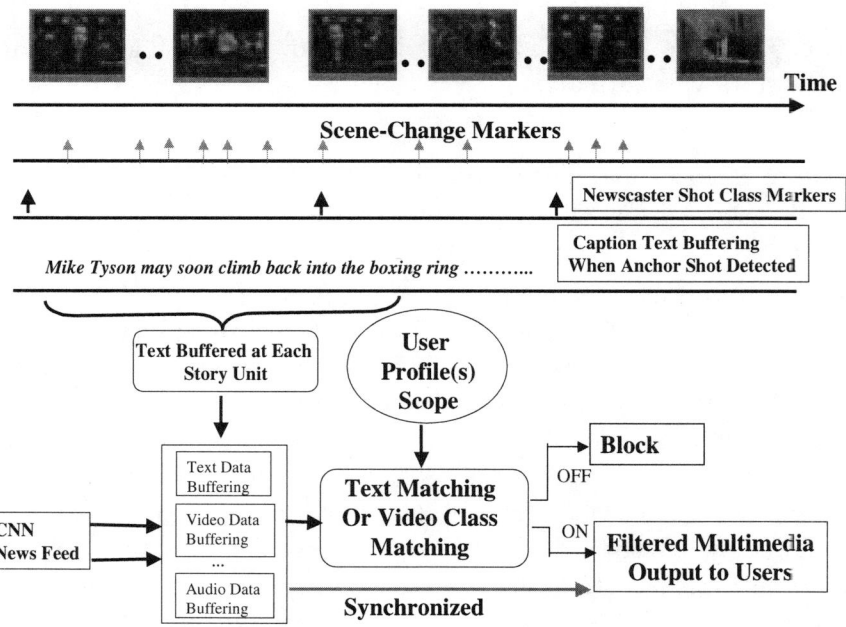

Figure 6.9. CNN news filtering.

6.5 Examples of System Usage

Having detailed the components that comprise on-line video content analysis, we now give two application examples to illustrate the functionality of the system and point out its strengths and weakness.

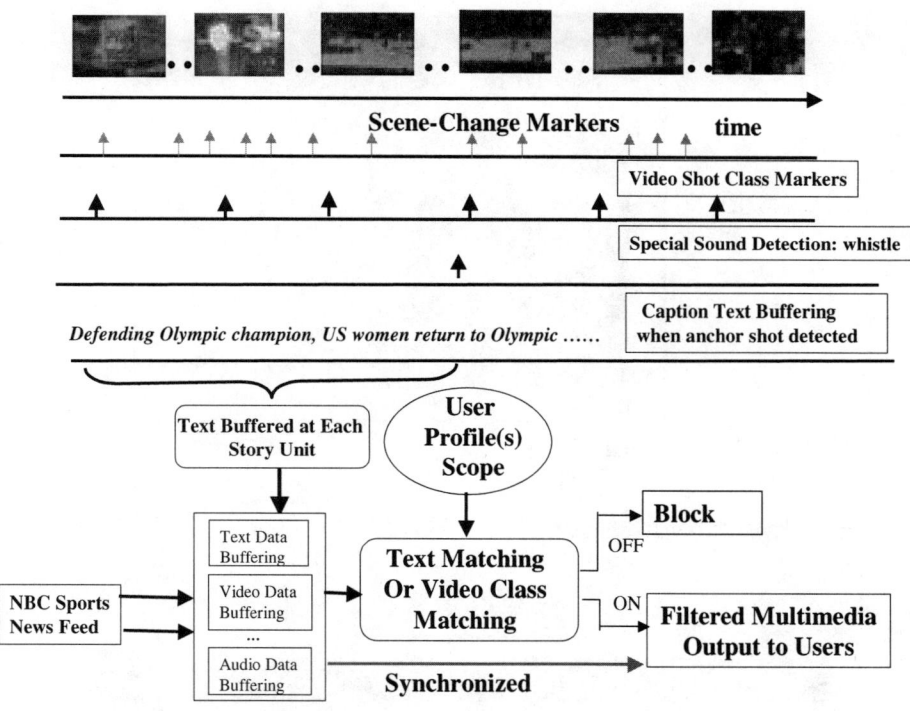

Figure 6.10. 2000 Olympic game event filtering.

6.5.1 On-Line CNN News Filtering

Figures 6.9 and 6.10 show data flows of on-line video filtering for CNN news video and NBC 2000 Olympic game event, respectively. Figure 6.11 shows the semantic multicast query interface. News topics can be specified by a user's profile for live video streaming. The query result window will display the media server address from which the requested video content is sent out. Real-time streaming tools, such as vic and vat, will then be launched automatically by listening to the appropriate server. Metadata analyzed by content agents can also be viewed by a Metadata Viewer.

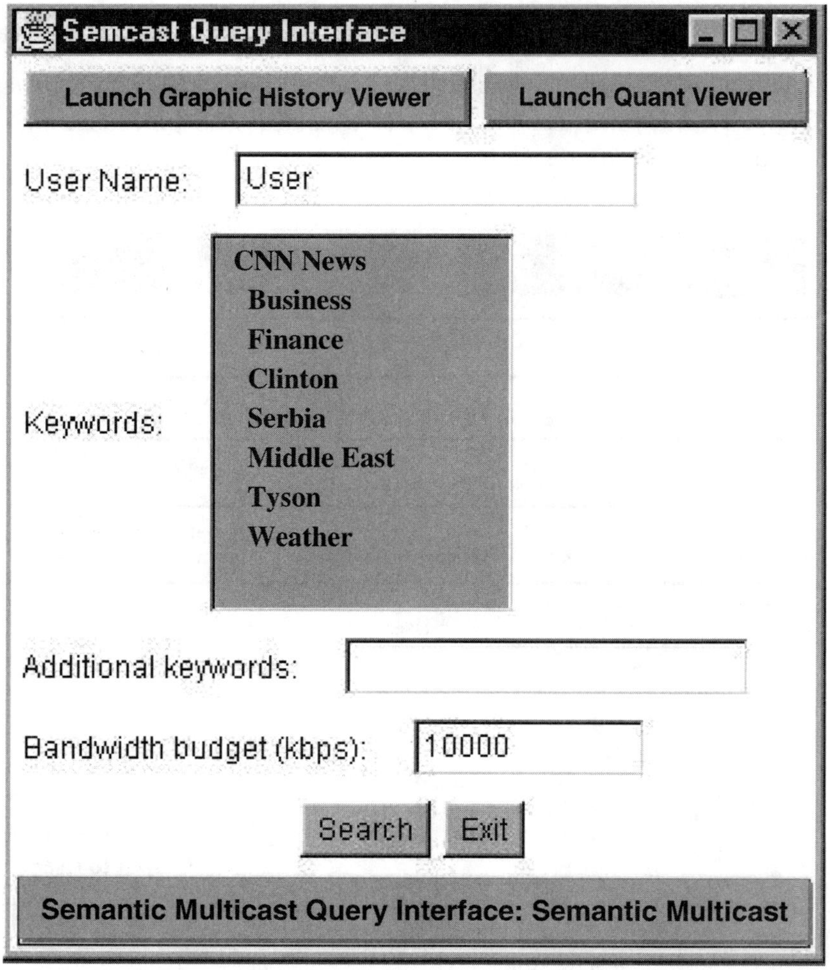

Figure 6.11. Semcast query interface.

We tested our system with a 30-minute CNN news video, of which two stories were related to President Clinton and one story is about the famous boxer Mike Tyson. The news video was automatically analyzed and organized as a concept tree which was managed by the video semantic classification/inference engine. In Figure 6.12, the user requests "Clinton," the query result shows the requested data locations are found, and the viewers only get real-time news videos related to President Clinton. Figure 6.13 shows data distribution to users who request both "Tyson" and "Clinton." For the whole 30 minutes of news content filtering, users can get the desired

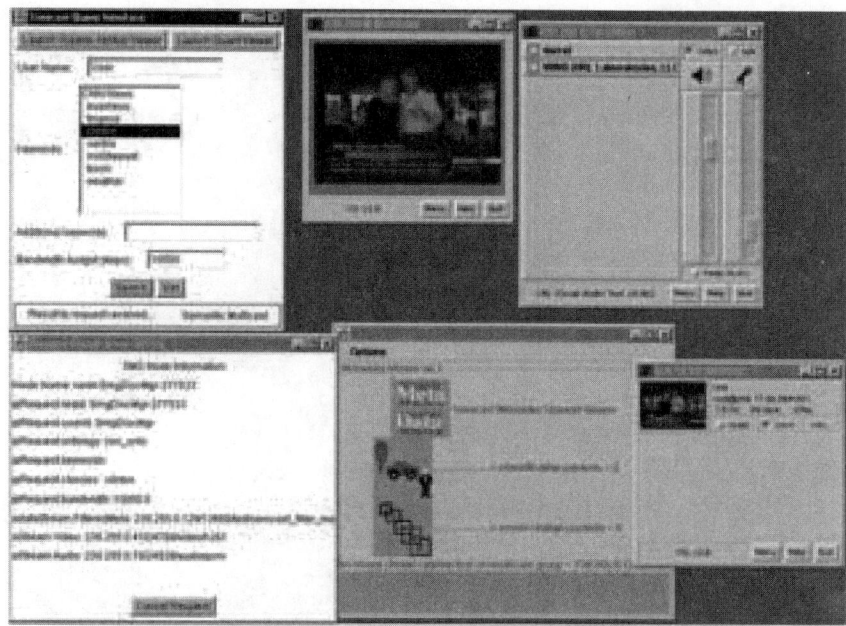

Figure 6.12. Filtering news that is related to President Clinton.

content specified on their profile based on the CNN hierarchy semantic concept tree with up to 100 percent accuracy. The speed of filtering agents in response to the increasing data rate is very good. As each agent generates a certain amount of latency when analyzing data, the overall latency for complicated data analysis and filtering would be noticeable but limited to less than one minute. We claim that the system is good enough even for emergency information alert and sharing.

6.5.2 Off-Line Video Database Query

A web-based query interface is shown in Figure 6.14 for off-line database queries. A concept tree is displayed in the interface, which can guide users in the query process based on the video hierarchical concept tree. Both key words and sentences can be submitted as queries. Figure 6.14 shows that when "basketball" is submitted in a query, video data and the corresponding caption text information will all be displayed in the result display interface. Then, users can go and pick the data they are truly interested in. Querying

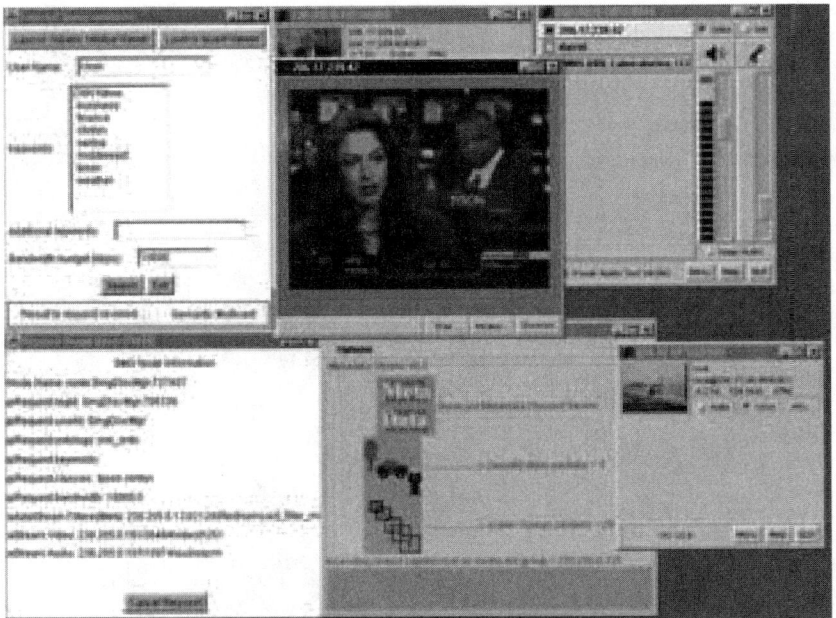

Figure 6.13. Filtering news that is related to Tyson.

with respect to a concept hierarchy is significantly more efficient and reliable than a search for specific key words, since views of collected data are refined as we go down the hierarchy.

6.6 Summary

Internet multicast clearly defines an essential layer of information delivery in collaborative applications. We presented the semantic multicast ([28]) that implemented a large-scale shared interaction infrastructure which provides a seamless environment for collecting, indexing, and disseminating the information produced in collaborative sessions. The content analysis agent infrastructure has been integrated into the semantic multicast framework seamlessly. Semantic multicast adds a level of interaction to the IP multicast to provide integrated support to partition and filter the information flows in order to enable effective dissemination and sharing of collaborative sessions over space and time (i.e., real-time and non-real-time) via the Internet and among a diverse user population.

Section 6.6. Summary   **155**

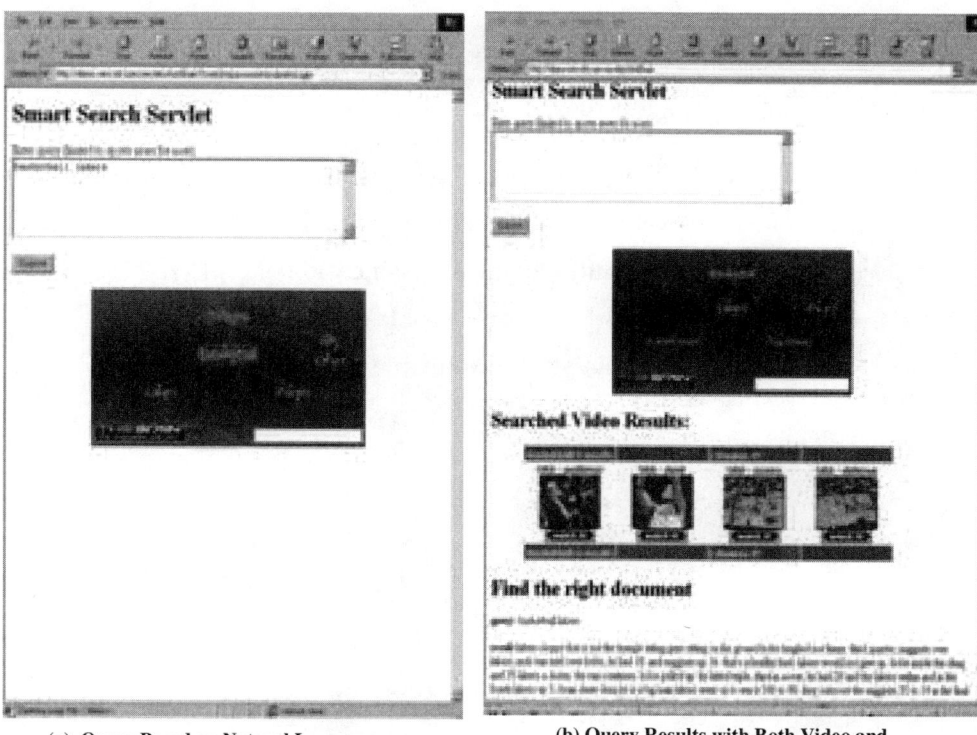

(a) **Query Based on Natural Language**   (b) **Query Results with Both Video and Caption Text Displayed**

Figure 6.14. Illustration of a web-based query interface.

An independent metadata channel was proposed to support communication and coordination among cooperative agents for efficient content analysis and filtering. The following are main research subtopics in content analysis agents being pursued in the development of the concept of semantic multicast and in the demonstration of its practical, scalable information-sharing and dissemination application system over the Internet.

- Fast content semantics extraction and correlation.
  Every stream must be processed in real-time (or near real-time) so that the stream can be properly matched and propagated in the semantic multicast graph. These efficient extraction techniques bootstrap the semantic multicast process so that streams are correctly forwarded to the most appropriate proxies for dissemination and further content

analysis.

- Effective and efficient tools implemented to support semantic multicast.
  These tools include:
    - on-line scene change detection applied to RTP-H.261 streams
    - on-line low-level feature extraction applied to RTP-H.261 streams
    - on-line key frame and sequence classification of RTP-H.261 streams
    - filtering tools based on user's profiles and
    - buffering and media stream multiplexing support.

Chapter 7

CONCLUSION AND FUTURE WORK

In this chapter, the summarization of this book is given in Section 7.1, the primary contributions are presented in Section 7.2, and suggestions for future work are highlighted in Section 7.3.

7.1 Summary of This Research

The subjects of this book are the design and implementation of an intelligent integrated system for on-line video segmentation and content analysis which provides a seamless environment for collecting, archiving, and disseminating multimedia information over the Internet, along with a fast and efficient architecture for the system as an integral part of the video dissemination and access infrastructure for "pull"-type applications. Multimedia information access tasks such as retrieval and filtering are the focus of this work.

The main body of the book started with an introduction to on-line video content analysis requirements and video content representations, followed by design and implementation of the system. Briefly, the system consists of three main modules. In the first module, fast on-line video segmentation algorithms were implemented and accomplished, and multicast video were automatically segmented into distinct scenes in real-time. In the second module, parts of the MPEG-7 compliable visual/motion/audio/text descriptors were implemented for further video content analysis. In the third module, which is the most important part of the system, an efficient and expressive rule-based and knowledge-based video classification system was proposed and implemented for video semantic content classification and analysis.

By using this system, the boundaries of video stories such as news or basketball games with different events were precisely set, and an accurate classification rate ranging from 70 percent to 90 percent was achieved. At the same time, knowledge-based video classifiers based on mixed media cues, video, audio, and text were established by supervised learning. The knowledge-based system consists of fact lists, contexts, and rules for hierarchical video semantic concept classification. The procedure is generic and can be applied to on-line multimedia dissemination and information access. Experiments were conducted on an efficient classification/filtering scheme for a CNN news video distributed over the Internet, and a rule-based semantic classification scheme for an NBC 2000 Olympic sports video program was also investigated and presented in this work. The system and the architecture were proved to be very effective and successful in both case studies.

7.2 Contributions of This Research

The major contributions of this book are summarized as follows:

- Proposal of a general on-line video analysis model.
  A multilayered general on-line video analysis model was proposed and it provided us with guidance to develop novel video content analysis algorithms and tools throughout the design and implementation of our system. The analysis model is general and it can be applied to any system which is related to video content analysis and access.

- Development of fast video parsing/segmentation algorithms for video disseminated by MBone Multicast.
  In this book, we solved the basic problem of on-line video parsing/segmentation, which is different from that of off-line algorithms, by fast video scene change detection algorithms. A detailed study of our scene change detection algorithms has been carried out under the MBone Multicast environment and it demonstrated superior performance on speed over off-line algorithms.

- Generation of a video summarization approach for time-sensitive on-line video streams.
  After scene change detection, a key frame for each scene was extracted to serve as a fast, condensed, and efficient summary of the whole video by on-line agents for the purposes of video browsing, indexing, and further analysis.

- Investigation of feature extraction schemes based on the nature of video signals and problems to be solved.

Signal-processing techniques including morphological, statistical, and clustering methods were applied to the representation and extraction of low-level perceptual video features.

- Proposal of a scalable video concept hierarchy tree scheme to abstract video semantic content.
  Video semantic content is more application dependent than perceptual content and thus not easily defined. The provision of multivalue-specified semantic content analysis is very important. Different from previous work on video content classification, which focused only on certain isolated levels of video semantic concepts, we proposed a more general video concept hierarchy tree to convey the multiple levels of semantics contained in the same video sequences. In this approach, a concept tree was represented by a hierarchical scheme consisting of different levels of video semantics, from which the relationship among different levels of concepts can be easily demonstrated. Classification rules for each level of classes were integrated into the concept tree scheme.

- Investigation of feature extraction/clustering techniques in video content analysis for semantic classification.
  A novel rule-based supervised video classification system was implemented for fast on-line video semantic content inference and classification. The relationship of video semantics classes and visual/motional feature descriptors was studied and learned by a supervised learning algorithm; this made optimal on-line feature extraction and filtering possible.

- Integration of mixed media cues from audio and text for video segmentation and semantic content analysis.
  Audio and text cues were integrated to facilitate video segmentation and content analysis. In our approach, various cues were learned using video content analysis rules. A text classifier effectively complements classification shortcomings and deficiencies of the hierarchical video/audio classifier on certain kinds of semantic concepts such as the team names for a game.

- Demonstration of a working on-line information filtering and query system.
  This research has provided solutions to agent-based system support,

architectural, and implementation issues by incorporating the following components:

1. integration of video content analysis infrastructure with mixed media cues
2. temporal synchronization of content with descriptions (synchronization with metadata, video/audio/caption text, etc.)
3. multiplexing of multiple video descriptions associated with a content item for transmission over an IP channel.

A comprehensive, intelligent system for on-line video content analysis and classifications has been designed and implemented. The study of its performance in terms of speed, accuracy, and functionality showed that it met the criteria and goals of the design and was among the most advanced techniques in terms of scope as well as quality, in achieving fast video information access over the Internet.

7.3 Future Work

There are a number of possible extensions and enhancements for the work presented in this book. They are given as follows:

First, we can identify new features and their descriptors for more effective video content analysis. There are two possible directions. One is to study the effectiveness of new features on specific video classifications while the other is to identify more effective new features and more efficient feature extraction methods for these feature descriptors. For example, besides statistical motion and color descriptors, we may also use features such as trajectory motion. One of the potential tasks is to search for the most effective descriptors to represent these new features. In addition, based on the experience built up with the supervised learning algorithm, we notice that most machine-learning algorithms depend very much on training feature vectors. For different feature vector sets, rules are different and the classification accuracy varies too. Thus, for the learning part, the more complete the feature vectors used in training, the greater the possibility to find the best descriptors and rules for the whole system's performance.

Second, it is interesting to perform further study on system performance with videos in different domains. We evaluated our system for filtering with on-line CNN news and for both filtering and querying with NBC Olympic game videos. Further studies on system performance, such as evaluating video semantic concept tree scalability, classification accuracy, and speed, may be helpful in making real application products.

Third, it is worthwhile to compare our proposed video classification methods with other classification algorithms. There are many other machine-learning algorithms used in the context of artificial intelligence, such as Bayesian networks, neural networks, and so on. Many researchers have also been working on model-based video classification. For example, Iyengar ([41]) used the hidden Markov models (HMMs) to classify image sequences. Because many events are time-progressive and HMM works very well for time-progressive pattern classifications, HMM might be a very good candidate to outperform the decision tree learning algorithm for such video events. Therefore, an important direction for future work based on this book is to study and compare the performance of our system by using different machine-learning algorithms based on performance measures such as classification accuracy and speed.

Finally, it is desirable to perform further study on combined classifiers with different media cues. We studied and implemented two classifiers (i.e., video/audio classifier and text classifier) with totally independent features and algorithms. The two different classifiers offer complementary information about patterns to be classified in the video concept tree, which could be harnessed to improve the performance of the selected classifiers. In the current implementation, heuristic rules are applied in choosing classifiers to optimize classification results. More theoretical and experimental work may continue in this direction. A possible solution is to use a Bayesian Network to train the media-cue optimizer in the proposed system to get combined classification results that outperform both of the classifiers. This also requires further research.

REFERENCES

[1] K.C. Almeroth and M.H. Ammar. The interactive multimedia jukebox project. URL:http://imj.gatech.edu/.

[2] E. Amir, S. McCanne, and H. Zhang. An application level video gateway. *Proceedings 4th ACM International Conference on Multimedia, San Francisco*, Nov. 1995.

[3] C. Apte, F. Damerau, and S.M. Weiss. Automated learning of decision rules for text categorization. *Transactions of Office Information Systems*, 12(3), 1994.

[4] F. Arman, A. Hsu, and M.Y. Chiu. Image processing on compressed data for large video databases. In *Proc. First ACM Int. Conf. Multimedia*, pages 267–272, Aug. 1993.

[5] F. Arman, A. Hsu, and M.Y. Chiu. Image processing on compressed data for large video databases. *Proceedings 1st ACM International Conference on Multimedia*, pages 267–272, Aug. 1993.

[6] F. Arman, A. Hsu, and M.Y.Chiu. Feature management for large video databases. In *Storage and Retrieval for Image and Video Databases*, pages 2–12. SPIE, 1993.

[7] S. Bibbs, C. Breiteneder, and D. Tsichritzis. Data modeling of time-based media. *D. Tsichritzis, ed., Visual Objects,Center Universitaire D'Inf oratique, Univ. of Geneva*, pages 1–22, 1993.

[8] L. Breiman, J.H. Friedman, R.A. Olshen, and C.J. Stone. *Classification and Regression Trees*. Wadsworth International Group, 1984.

[9] W. Buntine. Learning classification trees. *Statistics and computing*, 2:63–73, 1992.

[10] M. Celenk. Colour image segmentation by clustering. *IEE Proceedings-E*, 138(5), Sept. 1991.

[11] M.T. Chan, Y. Zhang, and T.S. Huang. Real-time lip tracking and bimodal continuous speech recognition. *IEEE Signal Processing Society 1998 Workshop on Multimedia Signal Processing, Los Angeles, California, USA*, Dec. 1998.

[12] S.F. Chang, W. Chen, H.J. Meng, H. Sundaram, and D. Zhong. Videoq: An automated content-based video search system using visual cues. *ACM Multimedia*, 1997.

[13] F. Chen, M. Hearst, D. Kimber, J. Kupiec, J. Pedersen, and L. Wilcox. Metadata for mixed media access. *Managing Multimedia Data: Using Metadata to Integrate and Apply Digital Data, McGraw-Hills*, 1997.

[14] Y. Cheng. Mean shift, mode seeking, and clustering. *IEEE Trans. Pattern Anal. Machine Intell.*, 17, 1995.

[15] T.S. Chua, S.K. Lim, and H.K. Pung. Content-based retrieval of segmented images. In *Proc. of ACM Multimedia 94*, San Francisco, Ca., 1994.

[16] W.W. Cohen. Learning trees and rules with set-valued features. In *AAAI-96: Proc. of the Thirteenth National Conference on Artificial Intelligence*, 1996.

[17] W.W. Cohen and Y. Singer. Context-sensitive learning methods for text categorization. *In Proc. of the 19th Annual ACM SIGIR Conference*, 1996.

[18] D. Comaniciu and P. Meer. Robust analysis of feature space: Color image segmentation. *Computer vision and Patter recognition*, June 1997.

[19] D. Crevier and R. Lepage. Knowledge-based image understanding system: A survey. *Computer Vision and Image Understanding*, 67:161–185, Aug. 1997.

[20] Son Dao, Eddie Shek, Asha Vellaikal, R. Muntz, L. Zhang, and M. Potkonjak. Semantic multicast: Intelligently sharing collaborative sessions. *Journal of ACM Computing Surveys*, Mar. 1999.

[21] A.M. Dawood and M. Ghanbari. Scene content classification from mpeg coded bit streams. *IEEE Signal Processing Society 1999 Workshop on Multimedia Signal Processing, Copenhagen, Denmark*, Sept. 1999.

[22] Y. Deng and B.S. Anjunath. Content-based search of video using color, texture and motion. *Proc. of IEEE Intl. Conf. on Imaging Processing*, 2:534–537, 1997.

[23] Y. Deng and B.S. Manjunath. Netra-v: Toward an object-based video representation. *IEEE Transaction on Circuits and Systems for Video Technology, Special Issue on Image and Video Processing for Interactive Multimedia Services*, 9 1998.

[24] Nevenka Dimitrova and Forouzan Golshani. Motion recovery for video content classification. *ACM Transactions on Information Systems*, 13(4):408–439, 1995.

[25] A. Divakaran, H.F. Sun, H. Kim, C.S. Park, and B.S. Manjunath. Report on the mpeg-7 core experiment on the motion activity feature. In *ISO/IEC WG11/M5030, MPEG-7 Proposal Evaluation Meeting, Vancouver*, 7 1999.

[26] N.D. Doulamis, S.D. Doulamis, and S.D. Kollias. A neural network approach to interactive content-based retrieval of video database. *International conference on image processing*, 1999.

[27] S. Eickeler and S. Muller. Content-based video indexing of tv broadcast news using hidden markov models. *International Conference on Acoustics, Speech and Signal Processing*, 1999.

[28] Son Dao et al. Semantic multicast: Intelligently sharing collaborative sessions. *Intelligent Collaboration and Visualization Program, BAA 97-09*, white paper, 1997.

[29] A.M. Ferman and A.M. Tekalp. Efficient filtering and clustering methods for temporal video segmentation and visual summarization. *J. Vis. Comm. and Image Rep.*, 9(4):336–351, Dec. 1998.

[30] M. Flickner, H. Sawhney, W. Niblack, J. Ashley, Q. Huang, B. Dom, M. Gorkani, J. Hafner, D. Lee, D. Petkovic, D. Steele, and P. Yanker. Query by image and video content: The qbic system. *IEEE Computer*, 28(9):23–32, Sept. 1995.

[31] Y. Gong, C.H. Chuan, and G. Xiaoyi. Image indexing and retrieval based on color histograms. *Multimedia Tools and Applications*, 2:133–156, 1996.

[32] Y. Gong, L.T. Sin, C.H. Chuan, H. Zhang, and M. Sakauchi. Automatic parsing of tv soccer programs. *IEEE Transactions*, pages 167–172, 1995.

[33] J. Hafner, H.S. Sawhney, W. Equitz, M. Flickner, and W. Niblack. Efficient color histogram indexing for quadtratic form distance functions. *T-PAMI*, 17(7):729–736, July 1995.

[34] A. Hampapur, R. Jain, and T. Weymouth. Digital video segmentation. *Proceedings 2nd ACM International Conference on Multimedia*, pages 357–364, 1993.

[35] Paul Heckbert. Color image quantization for frame buffer display. *SIGGRAPH Proceedings*, page 297, 1982.

[36] N. Hirzalla and A. Karmouch. Detecting cuts by understanding camera operations for video indexing. *Journal of Visual Languages and Computing, No. 6*, 1995.

[37] W. Holfelder. Mbone vcr - video conference recording on the mbone. In *Proc. of ACM Multimedia '95*, pages 237–238, San Francisco, Nov. 1995.

[38] J. Huang, Z. Liu, Y. Wang, Y. Chen, and E.K. Wong. Integrating of multimodal features for video scene classification based on hmm. *IEEE Signal Processing Society 1999 Workshop on Multimedia Si gnal Processing, copenhagen, Denmark*, Sept. 1999.

[39] Q. Huang, Z. Liu, A. Rosenberg, D. Gibbon, and B. Shahraray. Automated generation of news content hierarchy by integrating audio, video, and text information. *International Conference on Acoustics, Speech and Signal Proc essing*, 1999.

[40] Q. Huang, A. Puri, and Z. Liu. Multimedia search and retrieval: New concepts, system implementation and application. *IEEE Transactions on Circuits and Systems for Video Technology*, 10(5):679–692, Aug. 2000.

[41] G. Iyengar and A. Lippman. Models for automatic classification of video sequences. *SPIE*, 3312:216–227, 1997.

[42] G. Iyengar and A. Lippman. Models for automatic classification of video sequences. *Proc. SPIE Multimedia Storage and Archiving Systems*, 3312:216–227, 1998.

[43] A. Jaimes and S.F. Chang. Model-based classification of visual information for content-based retrieval. *Storage and Retrieval for Image and Video Database VII,SPIE99*, Jan. 1999.

[44] A.K. Jain and A. Vailaya. Image retrieval using color and shape. *Pattern Recognition*, 29(8):1233–1244, 1996.

[45] S. Jeannin. Mpeg-7 visual part of experimentation model. In *ISO/IEC WG11/N2822, MPEG-7 Proposal Evaluation Meeting*, Vancouver, Jul. 1999.

[46] Ki-Young Jeong, Keechul Jung, Eun Yi Kim, and Hang Joon Kim. Neural network-based text location for news video indexing. *International conference on image processing*, 1999.

[47] D.B Johnson, R.K. Taira, W.S. Zhou, D.R. Aberle, and J.G. Golden. Hyperad: Augmenting and visualizing free text radiology reports. *Radiographics*, 18:507–515, 1998.

[48] Kanai. Image segmentation using intensity and color information. *visual Communications and Image Processing*, Jan. 1998.

[49] M.S. Kankanhalli, B.M. Mehtre, and J.K. Wu. Cluster-based color matching for image retrieval. *Pattern Recognition*, 29(4):701–708, 1996.

[50] R. Kasturi and R. Jain. Dynamic vision. In R. Kasturi and R. Jain, editors, *Computer Vision: Principles*, pages 469–480. IEEE Computer Society Press, Washington, D.C., 1991.

[51] R. Korfhage. *Information Storage and Retrieval*. New York: Wiley, 1997.

[52] Z. Liu, J.C. Huang, and Y. Wang. Classification of tv programs based on audio information using hidden markov model. *IEEE*, pages 27–33, June 1999.

[53] M. Maybury, A. Merilino, and J. Rayson. Segmentation, content extraction and visualization of braodcast news video using multistream analysis. *Proceedings of the 1997 AAAI Spring Symposium on Intelligent Integration and Use of Text, Image, Video, and Audio Corpora, (Stanford University, California), Technical Report SS-97-03*, pages 102–112, Mar. 1997.

[54] S. McCanne and V. Jacobson. Vic: A flexible framework for packet video. *ACM Multimedia*, 1995.

[55] J. Meng, Y. Juan, and S. Chang. Scene change detection in a mpeg compressed video sequence. *In Proc. of SPIE*, 2419:14–25, Feb. 1995.

[56] T. Mitchell. *Machine Learning*. New York: McGraw-Hill, 1997.

[57] MPEG-Group. Applications for mpeg-7. *ISO/IEC/JTC1/SC29/WG11 /N2462*, Oct. 1998.

[58] MPEG-Group. Mpeg-7 requirements documents. *ISO/IECJTC1/SC29 /WG11/N2461*, Oct. 1998.

[59] A. Nagasaka and Y. Tanaka. Automatic video indexing and full-motion search for object appearances. In *Proc. IFIP TC2/WG2.6 Second Working Conf. on Visual Database Systems*, pages 113–127, 1991.

[60] A. Nagasaka and Y. Tanaka. Automatic video indexing and full-video search for object appearances. In E. Knuth and L.M. Wegner, editors, *Visual Database System II*, pages 113–127. Elsevier, 1992.

[61] Y. Nakajima, Y. Lu, M. Sugano, A. Yoneyama, H. Yanagihara, and A. Kurematsu. A fast classification from mpeg coded data. *International Conference on Acoustics, Speech and Signal Processing*, 1999.

[62] Y. Nakamura and T. Kanade. Semantic analysis for video contents extraction, spotting by association in news video. In *Proc. of ACM Multimedia '97*, Seattle,WA, Nov. 1997.

[63] Y. Nakamura and T. Kanade. Spotting by association in news video. *Proceedings of the 1997 AAAI Spring Symposium on Intelligent Integration and Use of Text, Image, Video, and Audio Corpora, (Stanford University, California)*, pages 113–119, Mar. 1997.

[64] A.M. Nazif and M.D. Levine. Low-level image segmentation:an expert system. *Pattern Analysis Machine Intellligence*, pages 555–577, Sept. 1984.

[65] N. Negroponte. *Being Digital*. Vintage Books, 1995.

[66] W.-T. Ooi, B. Smith, S. Mukhopadhyay, H.H. Chan, S. Weiss, and M. Chiu. The dali: A multimedia software library. In *SPIE Multimedia Computing and Networking*, San Jose, CA, Jan. 1999.

[67] F. Oppini and R. Leonardi. Audiovisual pattern recognition using hmm for content-based multimedia indexing. *9th International Packet Video Workshop*, 1999.

[68] K. Otsuji and Y. Tonomura. Projection detecting filter for video cut detection. In *Proc. First ACM Int. Conf. Multimedia*, pages 251–257, Aug. 1993.

[69] Y.C. Park, P.K. Kim, F. Golshani, and S. Panchanathan. Conceptualization and ontology: Tools for efficient storage and retrieval of semantic visual information. *Proceedings of SPIE Internet Multimedia Management Systems*, 4210:37–48, 2000.

[70] P. Parnes. The multicast media-on-demand system. URL: http://www.cdt.luth.se/~peppar/progs/mMOD/.

[71] N.V. Patel and I.I. Sethi. Video classification using speaker identification. *SPIE*, pages 218–224, 1997.

[72] R.W. Picard. A society of models for video and image libraries. *IBM Systems Journal*, 35:292–312, 1996.

[73] M.F. Porter. An algorithm for suffix stripping. *Program*, 14(3):130–137, 1980.

[74] J.R. Quinlan. Induction of decision trees. *Machine Learning*, pages 81–106, 1986.

[75] J.R. Quinlan. *C4.5: Programs for machine learning*. Morgan Kaufmann, 1993.

[76] V. Rijsbergen. *Information Retrieval*. Butterworths, 1979.

[77] Y. Rui, T.S. Huang, and S.-F. Chang. Image retrieval: Current technologies, promising directions, and open issues. *Journal of Visual Communication and Image Representation*, 10(1):39–62, Mar. 1999.

[78] G. Salton. *Automatic Text Processing: The Transformation, Analysis, and Retrieval of Information by Computer*. Reading, Massachusetts: Addison-Wesley, 1989.

[79] G. Salton and C. Buckley. Term weighting approaches in automatic text retrieval. *Information Processing and Management*, 24(5):513–523, 1988.

[80] G. Salton, W. Wong, and C.S. Yang. A vector space model for automatic indexing. *Communications of the ACM*, 18:613–620, 1975.

[81] D.D. Saur, Y.P. Tan, S.R. Kulkarni, and P.J. Ramadge. Automated analysis and annotation of basketball video. *SPIE*, 3022, Sept. 1997.

[82] H. Schulzrinne, S. Casner, R. Frederick, and V. Jacobson. Rtp: A transport protocol for real-time applications. *RFC 1889*, Jan. 1997.

[83] I.K. Sethi and N. Patel. A statistical approach to scene change detection. In *Storage and Retrieval for Image and Video Databases III, 2419*, pages 329–338. SPIE, Feb. 1995.

[84] B. Shabraray. Scene change detection and content-based sampling of video seqences. *Proc. SPIE*, 2419, 1995.

[85] D. Sharvit, J. Chan, H. Tek, and B.B. Kimia. Symmetry-based indexing of image database. *Proceedings IEEE Workshop on Content-based Access of Image and Video Libraries*, June 1998.

[86] S.W. Smoliar and H. Zhang. Content-based video indexing and retrieval. *IEEE Multimedia*, pages 62–72, Summer 1994.

[87] M.V. Spinivasan, S. Venkatesh, and R. Hosie. Qualitative estimation of camera motion parameters from video sequences. *Pattern Recognition Society*, 30(4):593–606, 1997.

[88] G. Sudhir, J.C.M. Lee, and A.K. Jain. Automatic classification of tennis video for high-level content-based retrieval. *IEEE Multimedia*, 1997.

[89] M. Swain and D. Ballard. Color indexing. *International Journal of Computer Vision*, 7(1):11–32, 1991.

[90] C.W. Therrien. *Decision Estimation and Classification: An Introduction to Pattern Recognition and Related Topics*. Wiley, 1989.

[91] Y. Tonomura, A. Akutsu, Y. Taniguchi, and G. Suzuki. Structured video computing. *IEEE Multimedia*, pages 34–43, Fall 1994.

[92] S. Uchihashi, J. Foote, A. Girgensohn, and J. Boreczky. Video manga: Generating semantically meaningful video summaries. *ACM Multimedia*, pages 308–392, Oct. 1999.

[93] N. Vasconcelos and A. Lippman. Bayesian modeling of video editing and structures: Semantic features for video summarization and browsing. *International conference on image processing*, 1998.

[94] H.D. Wactlar, M.G. Christel, Y. Gong, and A.G. Hauptmann. Lessons learned from building a terabyte digital video library. *IEEE Computer Mag.*, 32:66–73, Feb. 1999.

[95] H.D. Wactlar, T. Kanade, M.A. Smith, and S.M. Stevens. Intelligent access to digital video: Informedia project. *IEEE Computer Mag.*, 29:46–52, May 1996.

[96] J. Wang, W.-J. Yang, and R. Acharya. Efficient access to and retrieval from a shape image database. *Proceedings IEEE Workshop on Content-based Access of Image and Video Libraries*, June 1998.

[97] E. Wold, T. Blum, D. Keislar, and J. Wheaten. Content-based classification, search, and retrieval of audio. *IEEE Multimedia*, 3(3):27–36, 1996.

[98] B.-L. Yeo and B. Liu. Rapid scene analysis on compressed video. *IEEE Transactions on Circuit and System for Video Technology*, 5(6):533–544, Dec. 1995.

[99] M.M. Yeung and B.-L. Yeo. Video visualization for compact presentation and fast browsing of pictorial content. *IEEE Transactions on Circuit and System for Video Technology*, 7(5):771–785, Oct. 1997.

[100] M.M. Yeung, B.-L. Yeo, and B. Liu. Time-constrained clustering for segmentation of video into story units. *Proc. Int. Conf. Pattern Recognition, Vienna*, pages 375–380, Aug. 1996.

[101] H. J. Zhang, A. Kankanhalli, and S. W. Smoliar. Automatic partitioning of full-motion vide. *ACM/Springer Multimedia Syst.*, 1:10–28, 1993.

[102] H.J. Zhang, Y. Gong, C.Y. Low, and S.W. Smoliar. Image retrieval based on color features: An evaluation study. *SPIE Digital Image Storage and Archiving Systems*, Oct. 1995.

[103] H.J. Zhang, C.Y. Low, and S.W. Smoliar. Video parsing and browsing using compressed data. *Multimedia Tools and Applications*, 1(1):89–111, March 1995.

[104] T. Zhang and C.-C. Jay Kuo. Content-based classification and retrieval of audio. *SPIE, Storage and REtrieval for Image and Video Databases*, 3461:432–443, 1998.

[105] D. Zhong and S.F. Chang. Video object model and segmentation for content-based video indexing. *IEEE International Symposium on Circuits and Systems*, June 1997.

[106] D. Zhong, H.J. Zhang, and S.-F. Chang. Clustering methods for video browsing and annotation. In *SPIE Conf. on Storage and Retrieval for Image and Video Databases IV*, volume 2670, pages 239–246, 1996.

[107] W. Zhou and C.-C. Kuo. Knowlege-based inference engine for online video dissemination. *SPIE Internet Multimedia Management System*, 4210, Nov. 2000.

[108] W. Zhou, Y. Shen, A. Vellaikal, and C.-C. Kuo. On-line scene change detection of multicast (mbone) video. *SPIE Conf. on Storage and Retrieval for Image and Video Databases IV*, Nov. 1998.

[109] W. Zhou, A. Vellaikal, and C.-C. Jay Kuo. Rule-based video classification system for basketball video indexing. *Proceedings ACM Multimedia 2000 Workshops*, Nov. 2000.

INDEX

"Interaction stream" multicast, 11
"Missed-detects,", 38
"Multicast-aware" collaboration tools, 25
"Pull" applications of multimedia, 4
"Push" applications of multimedia, 4, 5
"Video Manga,", 4

Acoustical features, 50, 53, 78, 112
Acoustic features, 5
Algorithm design:
    goal of, 34
    scene change detection, 34
Annotation proxy, 148
Annotation proxy agent, 12, 19, 134
Area, 61
Audioconferencing, 29
Audio features representation and analysis, 70
    frequency-domain features, 71
        centroid frequency and bandwidth, 72
        energy ratio of some sensitive subbands, 72
        peak distribution of entire audio clip, 72
        peak of spectrum on each frame, 72
        short-time fundamental frequency, 71
        short-time fundamental frequency distribution for entire audio clip, 71
    time-domain features, 70–71
Automatic Annotations, 12
Automatic scene change description tags and key frames, 26

B-frames, 114, 116, 117
Bag-of-words, 73, 93
Bayesian methods, 87
Bayesian network classifier, 80
Bayesian Networks, 102
Bidirectionally predicted (BIP) type macroblocks, 116–117
Block aging algorithm, 31–32
Boolean vectors, 75
Booming camera operation, 63
Browsing, 3–7, 14
Buffering, 34, 35, 37, 39–41

Camera motion/motion flow cluster descriptor, 62–64
Camera motion types, 63

173

Camera operation, 62
Camera operations, 63
    booming, 63
    dollying, 63
    fixed panning, 63
    rolling, 63
    tilting, 63
    tracking, 63
    zooming, 63
CART algorithm, 87
CBR (Content-Based Retrieval) systems, 53
Center of object, 61
Centroid frequency and bandwidth, 72
Classification execution plan, 98–101
    knowledge-based content matching, 99
    Preprocessing for Feature and Feature Extraction Selection, 99
    video dissemination based on video classification, 100
Clustering, 50–51, 59
    color, 58, 59
        LUV color space, 59
        RGB color space, 60
    in feature space, 50–51
CNN, 2, 8, 14, 23, 24, 38, 81, 84
Collaborative Semantic Multicast, 9–12
    "interaction stream" multicast, 11
    modes of operation, 11–12
    semantic multicast system, goal of, 12
    underlying basis behind, 10–12
Collaborative sessions, 18
Color clustering, 52–60
Color descriptors, 49–60
    color clustering, 52–60
    color clusters, 52
    color histogram descriptor, 52–55
    color segmentation, 58–59
    dominant color, 58
    dominant color descriptor, 52, 56, 56 − −58
    spatial color, 52
Color histogram descriptor, 52–55
Color segmentation, 58–59
Conceptual Model Layer, layered video analysis model, 16
Conditional replenishment, 31–32
Content-based navigation, 1
Content-based queries, 7
Content-based video indexing, 3, 24
Content-based video indexing and retrieval systems, 3
Content agent coordination, 142
    communication between the proxy and the service assigner, 143
    metadata channel, 144–147
Content agent coordinationind, 147
Content agents
    buffer management, 137
    synthesis of multiple low-level features associated with a content item, 137
Content agents:
    annotation function, 136
    archival function, 137

Index

buffer management, 137
filtering, 137
synthesis of multiple low-level features associated with a content item, 137
temporal synchronization of content with descriptions, 137

Correct detects, 38

Dali Library, 51
data disssemination model, 16
Data scaling, and multicasting, 9
DCT images, 67
Decision tree learning algorithm, 80, 86, 92, 93
Dimensionality curse, 54
Dimensionality reduction techniques, 75–77
Directional characteristics, 69
Document frequency, 75, 94
Dollying camera operation, 63
Dominant color descriptor, 56–58
Dominant color extraction method, 57–59

Eccentricity, 62
Edge descriptors, 60
Edge detection, 59–60, 120
Edge patterns, 60, 119
Energy ratio of some sensitive sub-bands, 72
Entropy-based inductive tree-learning algorithm, 23
Execution Plan Generator (EPG), 100, 101
Experimental results, 104–129
CNN news video segmentation and indexing, 104–108

hierarchical NBC video classification, 126–129
hierarchical video classification on NBC 2000 Olympic video, 108–129
hierarchical video classification on NBC 2000 Olympic video, 108–129
low-level feature extraction, 111–122
audio feature extraction, 120
compressed domain descriptors, 116
motion descriptors, 114
text feature extraction, 122
on-line CNN news segmentation using scene classifications, 106, 107
program segmentation of NBC sports video, 109, 110
rule-Based video event classification for a basketball video, 122–126
sports event units segmentation, 111
table-of-contents generation for CNN news indexing, 107, 108
TV news video production rules, 104, 105

Expert systems, 81

FDY algorithms, 36, 42, 46, 47
FDYUV algorithms, 36, 46, 47
Feature and Content Layer, layered video analysis model, 16–17
Feature space, clustering in, 50–51
Filtering, 4–5

Filtering proxy, 134, 148, 150
    inputs to, 150
    outputs to, 150
First-order derivative masks, 60
Fixed panning camera operation, 63
Focus-of-expansion (FOC), 64
Focus-of-expansion (FOE), 64
Forward predicted (FWD) type macroblocks, 116–117
Frequency-based vectors, 74
Frequency-domain features, 71–73
    centroid frequency and bandwidth, 72
    energy ratio of some sensitive subbands, 72
    peak distribution of entire audio clip, 72
    peak of spectrum on each frame, 72
    short-time fundamental frequency, 71
    short-time fundamental frequency distribution for entire audio clip, 71
FSY algorithms, 36, 47
FSYUV algorithms, 36, 47
Fuzzy k-means algorithm, 51

General concept tree, illustration of, 84
General conceptual modeling of video, 5–6
Goals of algorithm design, 34

H.261 format, 27
Hidden Markov Model, 8, 130
Hidden Markov Model classifier, 80

Hierarchical NBC video classification, 126–129
Hierarchical video classification on NBC 2000 Olympic video
    Olympic video, 108–129
    hierarchical NBC video classification, 129
    low-level feature extraction, 111–122
    program segmentation of NBC sports video, 109
    rule-Based video event classification for a basketball video, 122
    sports event units segmentation, 111
    rule-Based video event classification for a basketball video, 129
Hierarchical video representation, 5–6
High-level semantic concepts, 15
High-level semantic meanings, 6
High-level video meanings, 22

I-frames, 114, 116–117
Indexing, video stream collections, 3
Information dissemination by multicast, 9
Information filtering, 1
Information gain, 80
Informedia project (Carnegie Mellon University, 80
Integrated services, 1
Intelligent content agents, 19
Intelligent multimedia presentation, 4

Intensity and spatial characteristics, 67–68
Interactive media services, 1
Interactive Multimedia Jukebox, 30
Interface, semantic query, 151
Interface to a multimedia repository, 135
Internet news directories, 2
Internet packet (IP) video, 25
Intra-H.261 algorithm, 30–32, 65
Intra-H.261 framing protocol, 35
Inverse document frequency, 75, 94
Ivs format, 27

k-means algorithm, 51
K-NN based classifiers, 80
Key frame clustering, feature extraction for, 119
Key frames, 24, 50
Knowledge-based classifier, 81
Knowledge-based content matching, 99
Knowledge-based systems, 81
  development of, 81
Knowledge-based techniques, 81
Knowledge-based video classifiers, 79
Knowledge-based video hierarchical classification, 78–103
  Bayesian network classifier, 80
  classification execution plan, 98–101
    knowledge-based content matching, 99
    preprocessing for feature and feature extraction selection, 99
    video dissemination based on video classification, 100
  decision tree learning algorithm, 80, 86–89
  flow chart of, 89
  Hidden Markov Model, 80
  Informedia project (Carnegie Mellon University), 80
  K-NN based classifier, 80
  knowledge-based systems, 81
  knowledge-based techniques, 81
  neural network classifier, 80
  on-line knowledge-based video classification system, 97, 102
    Media Browser (MB), 97
    Media Matcher (MM), 97, 101–102
    Media Planner (MP), 97–101
    Media Repository (MR), 97
  research, 79–93
  rule-based knowledge base construction, 89–96
    video classification rules in the knowledge base, 89–92
    video classifier with text cues, 93–96
    video classifier with video/aduio cues, 92–93
  video semantic concept tree, 78, 82–86
Knowledge Layer, layered video analysis model, 16
Knowledge Tree, 18

Laplacian mask, 60, 120
Latency time, 39–47
Layered video analysis model, 15–18
  Conceptual Model Layer, 17
  example of, 16
  Feature and Content Layer, 17

Knowledge Layer, 18
layers, 15–16
Raw Data Layer, 16
Video Segmentation Layer, 17
Learned decision trees, construction of, 98
Link-list buffer, 35–36
Low-level video features, 1, 15, 17
video classification based on, 7
Low-level visual features, 49
LUV color space, 59

Machine-learning tools, 22
Matching/querying process, 62
mb, 29
MBONE, 9, 27–29
MBone tools, 29–30
Interactive Multimedia Jukebox, 30
MBone VCR, 29
mMOD, 30
Media Concept Hierarchy (MCH), 97, 99, 100
Media Cues Optimizer, 102
Media Matcher (MM), 101–102
Media Planner (MP), 97–101
Metadata channel, 144–147
metadata ID, 146
packet length, 147
payload, 147
timestamp, 147
Metadata Viewer (MV), 149–151
Mixed media access to multimedia databases, 7–8
mMOD, 30
Moment-based shape features, 61
Motion activities descriptor, 65–70
direction of activity, 66
extraction methods, 67–70

directional characteristics, 69
intensity and spatial characteristics, 67–68
motion vector-based parameters for a shot, 69–70
shot activity histogram, 70
intensity of activity, 65
spatial distribution of activity, 66
temporal distribution of activity, 67
Motional features, 53
camera motion/motion flow cluster descriptor, 62
object motion descriptor, 62
representation and analysis, 62, 70
Motion descriptor extraction, direction of, 114
motion flow cluster descriptor, 62
Motion JPEG, 22, 27
Motion magnitude calculation, 115–116
MPEG-7, 3, 49, 65
Multicasting, 9
collaborative semantic multicast, 9–12
Multicast video, on-line scene change detection of, 25–48
space, see Scene change detection
Multimedia Content Description Interfac, see MPEG-7
Multimedia database retrieval, 1
Multimedia information access, challenges of, 2–5
Multistream media, videos as, 2
Multiword terms, 77

MY algorithms, 38, 42, 47
MYUV algorithms, 38, 42, 47

NBC 2000 Olympic video, 108–129
    hierarchical NBC video classification, 126–129
    low-level feature extraction, 111–122
    program segmentation of NBC sports video, 109
    rule-based video event classification for a basketball video, 122–126
    sports event units segmentation, 111
NBC Olympic game news, 24
Neural network classifier, 80
Normalized area, 61
Normalized center of object, 61
nv format, 27

Object motion activities, 62
Object motion descriptor, 62, 64–70
Object orientation, 61
Object trajectory descriptor, 62
Olympic video, 109–131
On-line knowledge-based video classification system, 97–102
    Media Matcher (MM), 101–102
    Media Planner (MP), 97–101
On-Line multimedia aontent analysis and annotations
    annotation dependency constraints, 14–15
    layers of annotation, concept of, 13–14
    real-time constraint , 13
    time granularity constraint, 14
    uncertainty of metadata accuracies, 14
    video semantic extraction constraint , 13
On-Line multimedia content analysis and annotations, 12–15
On-line multimedia data, increase in, 2
On-line scene change detection and key frame extraction, 137–142
    knowledge-based video classification, 141–142
    low-level feature analysis and extraction, 139–141
On-line video content analysis implementation, 9, 136–137
    annotation, 136
    archival funcation, 137
    buffer management, 137
    filtering, 137
    synthesis of multiple low-level features associated with a content item, 137
    temporal synchronization of content with descriptions, 137
On-line video content analysis infrastructure, 15, 18–20
On-line vs. off-line annotations, 26

P-frames, 114, 116, 117
Packet buffering, 34, 35, 37, 40
Packet buffering mechanism, 40
Parameter motion descriptor, 62
Parameter trajectory, 62
PDY algorithms, 37, 38, 47
PDYUV algorithms, 37, 38, 47
Personalized television services, 4

Prewitt and Robinson three-level masks, 60, 120

QBIC, 129
Query interface, 151

rat, 29
Raw Data Layer, layered video analysis model, 15
Real-time Internet video streaming, 7
Research:
    contribution to, 158–160
    future work, 160–161
    summary of, 157–158
RGB color space, 60
Rolling camera operation, 63
RTP (Real Time Protocol), 27–29, 34, 37, 38, 145
    header, 27–28, 35
    RTCP, 27
    rtpdump tool, 138
    rtptools, 28, 38
    unreliability of, 28
rtpplay tool, 138
Rule-based knowledge base construction, 89–96
    document frequency, 94
    inverse document frequency, 94
    term frequency, 94
    term frequency × inverse document frequency, 94
    video classification rules in the knowledge base, 89–92
    video classifier with text cues, 93–96
    video classifier with video/aduio cue, 92–93

Rule-based video event classification for a basketball video, 122–126
    hierarchical NBC video classification, 126–129
Rulebased classification system by supervised learning, 22–23

Scene change detection, 25–48
    algorithm design, 34–35
        and scene change detection accuracy and recall rate, 34
        goal of, 34
    algorithms/implementation, 35–38
    experimental results and discussion, 38–47
    frame sampling, 36
    full frame scene change detection, 36
    histogram comparison, 39
    MBONE tools, 29–30
    mixture method, 38
    partial decompression algorithms, 37–38
    pixel value comparisons, 33
    RTP (Real Time Protocol), 27–29, 34, 38
    shot boundary detection, 33
    vic and Intra-H.261 algorithm, 30–32
Scene changes, 21
Second-order Laplacian mask, 120
Segmentation proxy, 148–150
Semantic, defined, 82
Semantic content extraction, 2
Semantic meaning expression, 5
Semantic meanings, 6

Semantic multicast system, 12
  goal of, 12
Semantic multicast system architecture, 133–136
  content agents, 134–135
  information discovery graph (IDG), 135
  multimedia repository, 134
  semantic multicast graph (SMG), 133
  contextural focus, realization of, 135–136
  service assigner, 134
Semantic Request Parser (SRP), 97, 100
Semantic video indexing, 1
Shape descriptors, 53, 58, 61, 62
  area, 61
  center of object, 61
  eccentricity, 62
  matchingprocess, 62
  moment-based shape features, 61
  normalized area, 61
  normalized center of object, 61
  object orientation, 61
Short-time audio volume, 70
Short-time Fourier transform, 71, 72
Short-time fundamental frequency, 71
Short-time fundamental frequency distribution for the whole audio clip, 71
Shot activity histogram, 70
Shots, 21
  motion vector-based parameters for, 69

Similarity search, 8
Speech-to-text transcription, 14
Sports video classifications, 7
Stopping and stemming algorithm, 93
Stop words, 76, 77
Streaming video, 4–5, 13
  packetized nature of, 13
Summarization, 3–4
System integration, 132–156
  content agent coordination, 142–147
  communication between the proxy and the service assigner, 143
  metadata channel, 144–147
  examples of system usage, 151–154
  off-line video database query, 153–154
  on-line CNN news filtering, 151–153
  on-line scene change detection and key frame extraction, 137–142
  knowledge-based video classification, 141–142
  low-level feature analysis and extraction, 139–141
  on-line video content analysis implementation, 136–137
  annotation, 136
  archival function, 137
  buffer management, 137
  filtering, 137
  synthesis of multiple low-level features associated with a content item, 137

temporal synchronization of content with descriptions, 137
semantic multicast system architecture, 133–136

Term, defined, 73
Term frequency, 74, 75, 94
Term frequency × inverse document frequency, 75, 94
Text-searching engines, 2
Text feature representation and analysis, 73–77
    Boolean vectors, 75
    dimensionality reduction techniques, 75–77
    frequency-based vectors, 74–75
    vector space representation, 74, 75
Text features, 53
TFIDF weighting, 74
Tilting camera operation, 63
time-domain features, 70
Tracking, 63
Training-set of video documents, 5
Traveling in the film field, 63
Two-stage matching scheme, 60

User-interfaces, 8
User agent driven media selection and filtering, 4

vat, 9, 29, 144
Vector space representation, 74, 75
vic, 9, 29–44, 65, 144
Video/audio/text feature representation and analysis, 49–77
    acoustical features, 53
    clustering, 50–51
    motional features, 53
    overall software structure, 51–53
    acoustical features, 53
    features, 52–53
    motional features, 53
    text features, 53
    visual features, 52–62
Video/image semantic contents, 1
Video:
    and collaboration data streams, 25–29
    as rich multimodel source, 49, 137
    as rich multistream source, 80
    multicasting of, 29
Video browsing, 3–4
Video classification, *see* Knowledge-based video hierarchical classification
Video classification rules in the knowledge base, 89–92
Video classifier with text cues, 93–96
Video classifier with video/aduio cue, 92–93
Videoconferencing, 4, 29, 31
Video content, 5–8
Video content analysis, 9
Video content analysis, flow chart of, 20
Video content analysis module, componets of, 21–23
Video content analysis structure, 23
Video content classification, 3
Video content concept classification, 3
Video content parsing, representa-

tion, and description, 6
Video data, 3–17
Video database indexing, 49
Video database indexing:
   using low-level features in, 22
   visual features, 49
Video database indexing and retrieval, 3
Video database management systems, 6
Video databases, 4
Video feature extraction, 23
Video filtering, 4
Video indexing, 1, 3, 7, 24
Video parsing and analysis techniques, 7
Videophones, 4
Video Segmentation Layer, layered video analysis model, 15
Video segmentation proxy:
   inputs to, 148–149
   output of, 149
Video semantic concept tree, 82–86, 97
   creating, 83–86
   illustration of, 78, 83, 84
   video semantic content, characteristics of, 79–82
Video semantic inference engine, 15
Video shot detection, 51
Video streaming and filtering, 4–5
Video structures, 7
Video structuring methods, 6
Video summarization and browsing, 3–4, 7
video understanding problems, traditional, 1
Visual features, 52–70

color descriptors, 52–60
   color clustering, 58–60
   color histogram descriptor, 53–55
   color segmentation, 58
   dominant color descriptor, 56–58
edge descriptors, 53, 60
shape descriptors, 53, 61–62
   area, 61
   center of object, 61
   eccentricity, 62
   matching/query process, 62
   moment-based shape features, 61
   normalized area, 61
   normalized center of object, 61
   object orientation, 61

wb, 9, 29, 144
Web-based query interface, 153, 154
Whistle detection, 141
Word stemming, 76

Yahoo, 2, 4, 84
YUV histogram, 46

Zero-crossing rate, 70, 71
Zooming, 63

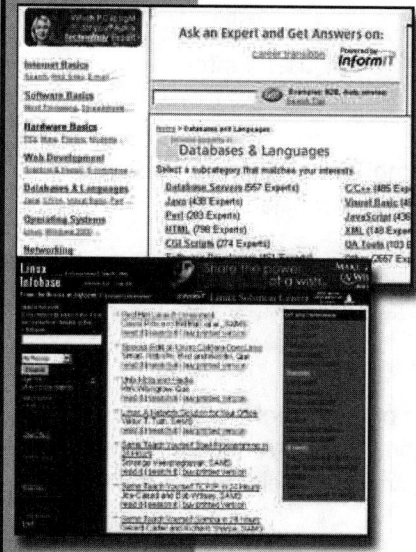

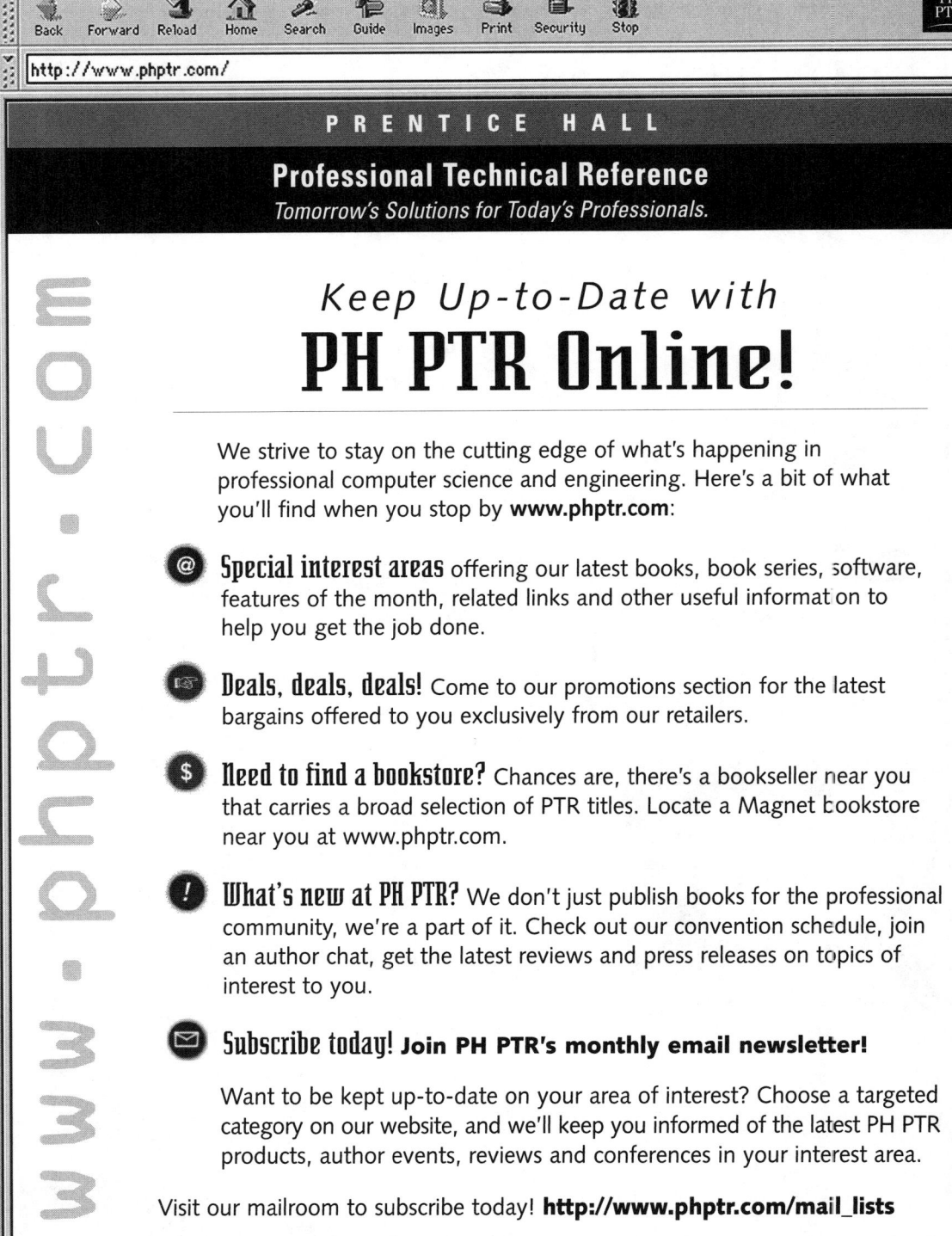